# SCIENCE
# IN THE
# LIGHT OF TORAH

# SCIENCE
# IN THE
# LIGHT OF TORAH

## A *B'Or Ha'Torah* Reader

edited by
**Herman Branover**
and **Ilana Coven Attia**

JASON ARONSON INC.
*Northvale, New Jersey*
*London*

This book was set in 11 pt. Goudy Oldstyle by Alpha Graphics of Pittsfield, New Hampshire, and printed by Haddon Craftsmen in Scranton, Pennsylvania.

**Library of Congress Cataloging-in-Publication Data**

Science in the light of Torah: A B'or Ha'Torah reader / edited by
    Herman Branover and Ilana Coven Attia
        p.   cm.
    Includes bibliographical references (p. ) and index.
    ISBN 1-56821-034-5
    1. Judaism and science.   2. Judaism—20th century.   3. Orthodox
Judaism.   I. Branover, Herman, 1931–   .   II. Attia, Ilana Coven.
BM538.S3B67   1994
296.3'875—dc20                                                    93-29371

Manufactured in the United States of America. Jason Aronson Inc. offers books and cassettes. For information and catalog write to Jason Aronson Inc., 230 Livingston Street, Northvale, New Jersey 07647.

# Contents

Contributors     ix

Preface     xv

### PART I
### GENERAL APPROACH TO LOOKING AT SCIENCE
### IN THE LIGHT OF TORAH

1 On Acknowledging the Creator in     3
Scientific Literature
*Aryeh Gotfryd*

2 The Religious Foundations of Science     11
*Avraham Kushelevsky*

3 The War between Religion and Science in the     15
Nineteenth Century and the Change in the
Twentieth Century
*Avraham Kushelevsky*

4 The Torah–Science Debates:     25
Some Random Thoughts
*Velvl Greene*

5  What We Cannot Know                                37
    *Paul Rosenbloom*


PART II

SPECIFIC APPLICATIONS OF LOOKING AT SCIENCE
AND ART DISCIPLINES IN THE LIGHT OF TORAH

Physics

6  The Creator in Creation                           51
    *Naftali J. Berg*

7  The Role of the Observer in *Halakhah* and        63
    Quantum Physics
    *Avi Rabinowitz and Herman Branover*

8  Geocentrism                                        81
    *Avi Rabinowitz*

9  Modern Physics and Jewish Mysticism              125
    *Gedaliah Shaffer*

Biology

10  Living Water: Concept of *Midah ke-neged*        135
    *Midah* and Cellular Homeostasis
    *Paul Goldstein*

Medicine and Public Health

11  Organ Transplantation and Jewish Law            141
    *Abraham S. Abraham*

12  Ethical Issues in Community Health              145
    *Velvl Greene*

# Evolution

13 The Evolutionary Doctrine 169
   *Lee M. Spetner*

14 Information Theory Considerations 179
   of Organic Evolution
   *Lee M. Spetner*

15 A Statistician Looks at Neo-Darwinism 185
   *Avraham M. Hasofer*

16 Torah, Science, and Carbon 14 197
   *Yaacov Hanoka*

# Philosophy

17 God and Rationality 207
   *Yitzchok Block*

18 Musings on (the Logic of) Repentance 225
   *George N. Schlesinger*

# Archaeology

19 Ancient Synagogues and the Temple 237
   *Asher Grossberg*

# History

20 Worlds of Difference 257
   *Yoseph Udelson*

# Music

21 Where the Roads Meet: Composer André 269
   Hajdu's Musical and Jewish Identity
   *Yaffa Goldstein*

Psychology

22  The Third Way: The Jewish Psychoanalytic          285
    Approach of Dr. Ida Akerman
        *Sarah ben-Arza*

Notes                                                                    295

Index                                                                    317

# Contributors

### Abraham S. Abraham, M.D., F.R.C.P.
Jerusalem-born, Abraham S. Abraham studied medicine in England, where he was elected a Fellow of the Royal College of Physicians in 1980. Currently Abraham directs the Department of Medicine B at the Shaare Zedek Medical Center in Jerusalem and is professor of medicine at the Hebrew University-Hadassah Medical School. Abraham's books on medicine and Jewish law, *Lev Avraham* (in two volumes) and *Nishmat Avraham* (in four volumes), are used throughout the world as authoritative guides. *Lev Avraham* has been translated into English under the title *Medical Ethics for Everyone*. An expanded and updated edition of this work has recently been published under the title *A Comprehensive Guide to Medical Halachah*.

### Sarah ben-Arza
A granddaughter of Rabbi Eliayu Ki-Tov, Jerusalemite Sarah ben-Arza studied musicology at Bar Ilan University.

### Naftali J. Berg, Ph.D.
Naftali Berg received rabbinical ordination in 1962 from the Rabbinical College of Canada. He received B.S.E.E. and M.S.E.E. degrees from the Illinois Institute of Technology in 1965 and 1966, respectively. In 1975 he completed his doctorate in Electrophysics at the University of Maryland. Since 1966 Berg has worked for the Harry Diamond Laboratories of the U.S. Army, where he specializes in three fields: the

transient nuclear effects on semiconductors; the interaction of light and sound in piezoelectric crystals; and the fusion of multiple sensors for target detection and tracking. He has received several patents and has authored over seventy technical papers and a book. Berg's technical work has been recognized by numerous awards, such as being named the Army Material Command Engineer of the Year in 1983. He is a driving force of the Habad movement in Baltimore, Maryland, where he lives with his wife and three children.

### Yitzchok Block, Ph.D.
Born in Nashville, Tennessee, Yitzchok Block holds a Ph.D. in Philosophy from Harvard University. He is a professor of philosophy at the University of Western Ontario in London, Canada, and his professional areas of interest are Aristotle and Wittgenstein. Block has published numerous papers and edited a book, *Perspectives on the Philosophy of Wittgenstein.*

### Herman Branover, Ph.D.
Born in Riga, Latvia, Herman Branover earned his Ph.D. from the Moscow Institute of Aviation in the then-new field of magnetohydrodynamics. After settling in Beersheba, Israel, in 1972 Branover created the Center for MHD Studies at the Ben-Gurion University. The Center includes teaching, research, and the development of an MHD energy generator. Chairman of the "Shamir" Association of Religious Academics from the former USSR in Israel, Branover also is the editor-in-chief of *B'Or Ha'Torah.* A recipient of the S. D. Bergman Prize for the development of new technology in Israel, Branover received the Knesset Speaker's Award in 1991 for his work for Russian immigrant absorption. He is a Foreign Member of the Academy of Science, Latvia; a full foreign member of the Russian Academy of Natural Sciences in Moscow; a member of the Moscow International Energy Club; and received honorary doctorates from the Russian Academy of Sciences and the Technical University of St. Petersburg.

### Paul Goldstein, Ph.D.
Born in Brooklyn, New York, Paul Goldstein is now a professor of genetics at the University of Texas at El Paso.

## Yaffa Goldstein

Born and educated in Israel, Yaffa Goldstein is a journalist living in Jerusalem.

## Aryeh Gotfryd, Ph.D.

A native of Toronto, where he still lives, Aryeh Gotfryd holds a Ph.D. in Ecology from the University of Toronto. Having authored and co-authored three books and many journal articles on ecological topics, Gotfryd now serves as an ecological consultant to government and industry and devotes himself to the Jewish community, as both teacher and troubadour. He edited *Fusion*, the proceedings of the first *B'Or Ha'Torah* Conference.

## Velvl Greene, Ph.D.

Velvl Greene was born and educated in Winnipeg, Canada, did his graduate work at Minnesota, and in the 1960s was one of the original participants in NASA's exobiology program, searching for life on Mars. He spent most of his academic career on the faculty of the University of Minnesota, teaching public health, doing research in environmental microbiology, and authoring nearly one hundred scholarly publications. In 1986 he made *aliya* to Israel with his family. Currently he is the Isaac and Elizabeth Carlin Professor of Public Health and Epidemiology at the Ben-Gurion University of the Negev in Beersheba and is also director of the Lord Jakobovits Center for Jewish Medical Ethics at the university.

## Asher Grossberg, M.A.

A native of Jerusalem, after studying at Yeshivat Kerem B'Yavneh, Asher Grossberg studied economics and business administration (M.A.) at the Hebrew University. Working as an economist at the Bank of Israel, Grossberg is also an accredited tour guide, specializing in Jerusalem and the Temple Mount during the Second Temple period.

## Yaacov Hanoka, Ph.D.

Yaacov Hanoka received his doctorate in solid-state physics from the Pennsylvania State University and now works as a senior physicist at the Mobil Tyco Energy Corporation in Waltham, Massachusetts. He has published dozens of papers in the fields of polytypism in crystals, solar cells, and electron beam–induced currents in solids.

### Avraham M. Hasofer, Ph.D.

Holding a Ph.D. in mathematical statistics from the University of Tasmania in Australia, Avraham Hasofer worked for two years at the Australian National University in Canberra as a research fellow on population genetics with Professor P. A. P. Moran, author of *The Statistical Processes of Evolutionary Theory*. From 1969 to 1992, he held the Chair of Statistics at the University of New South Wales. Hasofer is the author of over seventy-five scientific papers (including articles on genetics in the *Journal of Theoretical Biology*). His main professional interest is the application of statistics to civil engineering. He is the co-inventor of the Hasofer-Lind reliability index, which is widely used in civil engineering design. Hasofer is currently living in Melbourne, Australia, and teaching statistics at LaTrobe University.

### Avraham Kushelevsky, Ph.D.

At age nine, Avraham Kushelevsky immigrated with his parents from the United States to Israel, where he studied at the Slobodka Yeshiva in B'nei Brak and the Hebron Yeshiva in Jerusalem. Kushelevsky went to England to study physics and received his doctorate from Salford University. Since his return to Israel, he has been teaching and conducting research in medical physics at the Ben-Gurion University of the Negev in Beersheba and chairs its nuclear engineering department.

### Avi Rabinowitz, Ph.D.

American born, Avi Rabinowitz comes from a rabbinic family and has an extensive yeshivah education. He completed a doctorate in theoretical physics at New York University, has taught at various universities in the United States and Israel, and has published articles in Jewish and physics journals. Currently engaged in postdoctoral research in physics at the Ben-Gurion University of the Negev in Beersheba, Rabinowitz is also working on a short commentary on the Torah as well as two manuscripts on the Creation account in Genesis.

## Paul Rosenbloom, Ph.D.

Paul Rosenbloom was born in Portsmouth, Virginia, and completed his doctoral work in mathematics at Stanford University. From 1965 until his retirement in 1990, he taught mathematics and mathematics education at Columbia University and its Teachers' College. Since 1990 he has been an adjunct professor of mathematics at York University in Ontario, Canada.

The author of six books and dozens of research papers, Rosenbloom was awarded the Fréchet Prize by the French Mathematics Society in 1950. In 1953–1954 and 1971–1972 he was a member of the Institute for Advanced Study. From 1954 to 1957 Rosenbloom served on the Advisory Panel of the National Science Foundation.

## George N. Schlesinger, Ph.D.

After ten years of *yeshivah* study, George N. Schlesinger studied science and became a professor of philosophy. Teaching at the University of North Carolina in Chapel Hill, Schlesinger has published over one hundred fifty papers and nine books. His latest books are *New Perspectives on Old-Time Religion* (Oxford University Press) and *Sweep of Probability* (Notre Dame University Press).

## Gedaliah Shaffer, M.A.

Gedaliah Shaffer grew up in Boston and received an M.A. in Theoretical Physics at Princeton University. Living now in Brooklyn, New York, he works as President of the Automated Information Systems and has published numerous papers in physics and mathematics journals.

## Lee Spetner, Ph.D.

Lee Spetner took his doctorate in physics at the Massachusetts Institute of Technology in 1950. From 1951 to 1970 he was engaged in research and development at the Applied Physics Laboratory of Johns Hopkins University, where he was a member of the Principle Professional Staff. Spetner has taught physics, computer science, and communication theory at M.I.T., Howard University, Johns Hopkins University, and the Weizmann Institute of Science. He is listed in *Who's Who in Science* and *Who's Who in the World*.

Spetner settled in Israel in 1970, where he was technical director of Eljim Ltd. Later he was manager of the Navigation and Wave
Propagation Division at Elbit Ltd., which he left in 1985 to do private
consulting. In 1990 he retired.

Since 1963, when he spent a year in the Biophysics Department
of Johns Hopkins University, Spetner has made a hobby of the study
of organic evolution. He has published several articles on the subject
in professional journals, including *Journal of Theoretical Biology, Transactions on Information Theory, IEEE Transactions on Information Theory,*
and *Nature.* He is currently preparing a book which shows, on the basis
of information theory, that the neo-Darwinian theory of evolution does
not work. He offers his own hypothesis of evolution to explain the facts
usually used to support evolution. His hypothesis does not invoke random events, as does the neo-Darwinian theory.

## Yoseph Udelson, Ph.D.

Yoseph Udelson is a specialist in modern Jewish and European intellectual history, as well as the history of television technology. Udelson's
study of Israel Zangwill explores the challenges modernity poses for
Jewish identity, while his book on the history of American television
technology is considered the authoritative history in this field. He
currently is teaching at Tennessee State University.

# Preface

For a number of generations, mankind has been lured by the self-aggrandizement of Western humanism to look at the Bible through a supposed "light" of science. The *Haskalah* movement took upon itself to transplant this misperception of the Enlightenment into Jewish minds. The Authorship of the Torah, the Creation of the universe, the very existence of our souls, and our accountability to God have been questioned under the pretense of scientific progress and enlightenment. This questioning resulted in the spread of atheism and nihilism; it converted boys fresh from yeshiva into Marxist revolutionaries and turned tender Jewish children into brutal fighters for an imagined universal social justice. Paradoxically, this erosion is continuing—even now, when Marxism is collapsing both in theory and in practice. Somehow, the idea still persists that "science makes us atheists." This arrogance breeds uncertainty and anxiety, the breakdown of family life and moral values, and a spiraling increase in drug abuse and violence.

The heirs of the spirit of questioning of the Age of Enlightenment, which led to mankind's most materialistic and cruel century, relentlessly refuse to acknowledge that pure science no longer fits into the popular ideology of one hundred to two hundred years ago. Physics is no longer deterministic (see Kushelevsky, Greene, Rosenbloom, Shaffer, Berg, Rabinowitz, and Rabinowitz and Branover in this volume), and evolution remains a metaphysical system of ideas unproven by empirical observation and quantitative analysis (see Spetner, Hasofer, and Hanoka in this volume).

In 1981, a close group of colleagues and I decided to publish our dissenting opinions on the prevailing attitude of Jews toward science and religion. What troubled us was not only the attack on the inner core of Torah-devoted Jews from the outside. More disturbing was the fact that so many insiders—Torah-educated and observant Jews—were taking an apologetic approach in the Torah–science argument. Mostly subconsciously but sometimes consciously, they were seeing the Torah outlook as inferior to science. They were trying to adjust the Torah interpretation to fit into the socially acceptable consensus, rather than looking upon science and popular attitudes from the Torah perspective.

My fellow dissenters and I were all senior scientists engaged in research and university teaching. Every one of us had examined the differences between the genuine science we worked in and the pseudoscience or "scientism" pervading the mass culture we lived in. We all had found our way to strict Torah observance through the teaching and personal guidance of the Lubavitcher Rebbe, Menachem Mendel Schneerson. The Rebbe instructed us how to overcome the dichotomies we faced. When writing to the Rebbe, we did not have to rephrase our university language (as is well known, the Rebbe holds higher degrees in physics, engineering, and philosophy from the Sorbonne, the University of Berlin, and the Paris Polytechnical Institute). Under the Rebbe's guidance, scientists such as Naftali Berg, Yitzchok Block, Velvl Greene, Yaacov Hanoka, Avraham Hasofer, Paul Rosenbloom, Gedaliah Shaffer, and I held endless discussions and sometimes even structured seminars to clarify different aspects of the so-called Torah–science conflict.

In 1981 we decided to express our gratitude to the Rebbe for his guidance by dedicating a Torah-and-science journal in honor of his eightieth birthday. He gave us his blessing and approved our suggested title—*B'Or Ha'Torah*. Meaning "in the light of the Torah," the name was adopted from *Torah Or*, a commentary on the books of Genesis and Exodus written by the first Lubavitcher Rebbe, Schneur Zalman of Liadi.

The title accurately describes the journal. We discuss science and popular misconceptions of it, the arts, and the spiritual crises of modern life itself in the light of the Torah. Our point of view, our measuring stick, and constant value come from the Torah. We believe that the scientific interpretation of empirically observed new phenomena should be analyzed by the Jewish scientist from a Torah perspective also. A contradiction between the two does not warrant rejecting the scientific interpretation without sufficient scientific evidence. However, the fact that a particular scientific interpretation contradicts the Torah should cause serious and permanent concern. Obviously, the conflict ultimately will be resolved through the advance of science in favor of the Torah because the Torah is the blueprint of the universe.

This approach allows us to express reservations about a certain theory or interpretation long before it is discarded on a purely scientific basis. A large-scale example of the advantage of taking precautions against ideas antithetical to the Torah is the rise and fall of the concept of universal determinism. From the age of Newton and the other founders of classical science to the beginning of the twentieth century, the concept of determinism was overwhelmingly accepted at all universities. Combined with the seemingly self-evident idea that the natural sciences can provide a relevant explanation and means of computation of all phenomena—including those related to life, human behavior, and even thought—the principle of universal determinism led Laplace to say that if given an exact description of the condition of every atom in the universe, the human mind could calculate and predict all natural events, including those of man.

Laplace's famous statement contradicts the fundamental Torah principle of free will. It also makes prayer meaningless. There is no place for free will if human decisions are determined by foreseeable electronic currents, chemical reactions, and so forth in the brain and body. If Laplace were right that everything is firmly, scientifically predetermined, it would be useless to pray for health, rain, or harvest. Therefore, universal determinism and Laplace's statement were rejected by Torah Judaism. Determinism was revoked at the beginning of this century by the development of modern science, in particular by quantum physics and now also by the new theory of chaos.

Modern science has abolished a number of other ideas that had long been assumed to show the "antiscientific essence" and "backwardness" of the Torah. Most significantly, modern science has learned to assess its own limitations. Modern science has revealed how infinitely complex and inexhaustibly intricate are the structure and functioning of nature. This is a far cry from the naive and simplistic views of classical science.

Even more important is the acknowledgment by many philosophers of science—primarily, Karl Popper—of the inevitable sharp turns and jumps in the development of scientific theories. According to Popper, science advances through attempts to refute previous theories. Therefore, Popper says, the mere idea of scientific truth is meaningless because a scientific theory, with all its grandeur and practical importance, is no more than the best possible, not yet refuted generalization and interpretation of the empirical findings.

The Popperian approach places the Torah–science debate in the right perspective. Although it is very significant that Einstein and Bohr were in greater agreement with the Torah than the nineteenth-century scientists, it is possible that this concurrence is only an isolated incident. Tragically, far-reaching damage can result when Jewish apologists adjust their approach to the Torah on the basis of a single turn of events in the ever-changing development of science.

Another example (although not in the natural sciences) of how "advanced" theories quickly become absurd anachronisms is Marxism and dialectical materialism. For me, as a former Soviet Jew, this example is of especially great importance. How many Jews abandoned the Torah for what they thought was a more just and enlightened social system! Today, after the failure of the seventy-year-long tragic experiment of the former USSR, who can defend the social advantages, the morality, or even the economic efficiency of the communist system? Who could have predicted the speed with which God's finger knocked down the massive Soviet empire? The papers by Avraham Kushelevsky and Velvl Greene in this volume, which touch upon the inevitable demise of communism—although correct in principle—are outdated today, only ten years after they were written.

Communism as a social and political system is collapsing, but ironically the materialistic atheism that conceived it is still thriving.

Materialism and atheism are blocking humanity from its source of vitality. *B'Or Ha'Torah* strives to let its readers understand what a mistake it is for a Jew to discard his identity and undermine his moral fiber in order to cater to the whims of social and intellectual fashion.

In thirteen years, fourteen volumes of *B'Or Ha'Torah* have appeared, including separate English and Hebrew issues. The list of contributing authors includes many names beyond the original nucleus of our discussion group. Some are *baalei teshuvah* (returnees to Judaism); others are religious from birth. Some are *Hasidim*, but many are not. Their common denominator is that they live by the law of the Torah. They look at scientific, social, and cultural phenomena from the perspective of the Torah. Belonging to the most highly creative and productive strata of modern society, they direct university and hospital departments, laboratories, and industrial companies. They participate in international conferences, write books, and win prestigious awards. But the unshakeable foundation of their existence is the Torah—learning it and living by its over three-thousand-year-old tradition. Their thought and belief system—the source of their strength—is presented on the following pages.

The present volume, the ongoing publication of the English and Hebrew editions of the *B'Or Ha'Torah* journal, and the international *B'Or Ha'Torah* conferences are possible first and foremost through the guidance of the Lubavitcher Rebbe and his constant loving attention to Torah-and-science matters and to the Torah-observant natural scientists working and writing in this discipline.

As already mentioned, the intitial *B'Or Ha'Torah* group has expanded manifold over the years. The contributing authors in this volume represent only a fraction of all the scientists currently involved. I would like to express my thanks to all of them for their dedication, for their readiness to generously contribute time and effort in spreading the light of knowledge to perplexed Jews. Working long, hard hours at universities, laboratories, and industrial firms, they are able to find time to study Torah-and-science problems usually only after midnight. Their sense of responsibility and enthusiasm, however, overcomes their fatigue.

Finally, I want to express my sincerest gratitude to someone whose selfless devotion and talent are the real cornerstone of the *B'Or Ha'Torah* journal. I mean, of course, Ilana Coven Attia, who combines in one person over a dozen positions ordinarily found in an editorial staff and publishing house.

Herman Branover
Jerusalem

# I

# General Approach to Looking at Science in the Light of Torah

# 1
# On Acknowledging the Creator in Scientific Literature

*Aryeh Gotfryd*

While publishing two recent articles,[1] I encountered editorial resistance to my using the Acknowledgment section to recognize and thank the Creator for His[2] contribution to my ecological research, a contribution which I felt was essential, relevant, and noteworthy. These experiences prompted me to compose this note in order to explain the validity of acknowledging the Creator in today's scientific climate.

## WHERE CREDIT IS DUE

There is a virtually universal custom among science authors to use the Acknowledgment section of each article to publicly thank selected agencies and individuals who helped bring the research to fruition. If it is legitimate to thank clerical staff, field assistants, esteemed colleagues, and funding agencies, then why not the Creator? After all, without His input there would be neither cosmos nor pondering researcher.

## MODERN PHYSICS CONCURS

This alone is enough justification for an acknowledgment because thanks and credit are due to the source of benefit. But many of course question whether there exists a transcendent Being who acts as a Source of benefit. Nonetheless, according to the new physics and the new cosmology, there is evidence that some sort of creation did occur and, furthermore, that the entire universe and its laws are guided and sustained by something above nature.[3]

Now whether the cosmic stage was set in 18 billion years or 5,748 years, or is continually renewed every moment, is irrelevant to the issue of acknowledging the Creator, because in each scenario, we too are created beings; whether directly or indirectly we too are effects of the First Cause. The Creator is not just commendable for the job He did on primordial hydrogen. Everything which came about since Creation is equally dependent on the Creator including human life and thought, and, therefore, thanks for current phenomena are quite in order.

## THEOLOGICAL OBJECTIONS ARE RED HERRINGS

One may argue at this point that the Creator doesn't need, want, or even hear our thanks, that He didn't do His work for our benefit, and that He is not involved with mundane events. Those are theological questions which are irrelevant to the issue of a science editor's responsibility to allow the author to acknowledge whoever made a noteworthy contribution to the work. The irrelevance of those arguments can be demonstrated by analogy to the case of a researcher who receives multifaceted support and guidance from a globally and thoroughly renowned expert in the field, an individual who does not need the thanks and who aided the researcher for his own purposes. It is difficult to conceive of any editor who, in such a case, would think twice about printing a brief statement of recognition or gratitude, even if the editor imagined that the famous expert had died. Similarly such arguments provide no justification for allowing anyone's particular theological opinions to suppress an acknowledgment of the Creator of the universe.

## CLARIFYING AXIOMS AND ASSUMPTIONS

An argument was recently raised by the editor of a noted ecological journal, in response to a statement that I made at the end of an article dealing with measurement bias in ecological data.[4] The editor viewed my acknowledgment of "the Creator as the only being having an unbiased point of view" as axiomatic, obvious to any thinking person, and therefore inappropriate for a journal. But was my statement axiomatic? The Oxford Dictionary as well as the logicians[5] defines an axiom as an established principle or self-evident truth. I personally know several thoughtful academics who would have happily challenged the truth of my statement (had it been published), and I think any random sample of today's research community would sport a significant number of dissenting opinions on various grounds (e.g., the followers of Kerns[6]). What is obvious to one is not necessarily evident to another. Since concepts of the Creator and His attributes are neither established nor self-evident to all or even nearly all, it seems incorrect to call my statement axiomatic, regardless of its truth.

Perhaps the editor was referring to another connotation of the term *axiom*, equating it with an assumption. If this is the case, what's wrong with stating relevant assumptions? On the contrary, the explicit statement of axioms, assumptions, and reasoning procedures may be just as important as detailing the experimental and analytical methods (although much more rarely done). Greater awareness of and stricter adherence to scientific axioms would help curb the unfortunate proliferation of antiscience within science.[7] Antiscience, or scientism, is the tendency to claim that unverifiable assumptions and conjectures are scientific facts and theories.[8]

## LIVING UP TO SCIENTIFIC CRITERIA

It is necessary to highlight the distinction between scientism and true science. Historically science has considered that (1) only a direct observation or measurement constitutes a scientific fact, and (2) only a theory disprovable by direct measurement may be considered scientific.[9] It is therefore axiomatic that if a theory cannot generate a critical, falsifiable hypothesis, it is nonscientific. If this axiom were stressed in biological literature, there would be less confusion about the status of many popular concepts, for instance, the belief that today's multitude of species arose from one or a few ancestors through the natural selection of randomly generated beneficial mutations. This belief would be recognized as a nonscientific theory since it cannot be falsified empirically.[10] As a result of this (among many other considerations),[11] the situation remains that "the facts of paleontology conform equally well with other interpretations, e.g., divine creation, . . . and can neither prove nor refute such theories."[12] Such statements are viewed, however, as heretical because prehistoric speciation through evolution has itself become an article of faith among scientists despite its flagrant nonconformity to a prior scientific axiom.

Other tenets of scientism, abstracted from Klahr,[13] include the beliefs that (1) everything can be explained rationally and any question which cannot be explained rationally is not meaningful; (2) all human behavior is deterministic, following natural laws of heredity, environment, and self-need; and (3) there are billions of planets in the universe with intelligent forms of life living on some of them. Although statements like these have some factual content, they are unverifiable, being neither facts nor scientific theories. They are merely fashionable opinions masquerading as science.

## ACKNOWLEDGING DIVINITY: PRECEDENTS FROM THE LITERATURE

It seems that part of the reason for the editorial resistance is that acknowledging the Creator in a legitimate scientific framework seems unconventional. Nonetheless many leading scientists, both in the past and recently, have managed to publish relevant comments about the Creator in the scientific literature, and not just in the Acknowledgments but in the very body of their articles and books.

For example, Redi, who pioneered controlled biological experimentation, stated that his faith in the command of the Supreme Creator motivated his famous experiments debunking the then-popular doctrine of spontaneous generation.[14] Descartes, too, derived his scientific principles not from empirical observation but from his metaphysics.[15] Accordingly, he saw God's immutability as the cause of the regularity of material behavior. Descartes, like Redi, empirically tested his ideas, but his natural laws themselves (which were qualitatively equivalent to the classical mechanical principle of inertia and the conservation of momentum in closed systems) stemmed from his belief that

> The first or primary cause of motion is God Himself who acts not only at the creation to impart motion to matter at the first instant, but who continues to act at each instant [by imparting motion to particles, and who] creates the universe anew at each instant.[16]

More recently, the renowned astronomer and mathematician Fred Hoyle, after challenging evolutionary dogma in a recent book, devotes an entire chapter to a discussion of God.[17] Schrödinger,[18] via a scholarly discussion on the nature of data, concludes that different observers would not concur unless our minds were parts of the Divine Mind. Hirsch[19] proposes a standard scientific notation to represent God in order to abstract the qualities of absolute uniqueness and boundlessness from any particular religious perspective. He presents the traditional mathematical symbolism, . . . $00^{00}$, and Adler's refined exponential infinity, . . . $00^{00}$ . . . , as legitimate functional descriptors, objectively alluding to the omnipotence, omnipresence, omniscience, eternity, and universality of the Almighty.

## MONOTHEISM AND SCIENCE: GOOD WORKING PARTNERS

The theory and practice of science are not at all incompatible with belief in the Creator.[20] On the contrary, the current tendency to divorce the Creator from our scientific understanding of nature removes the very *raison d'être* for pure scientific inquiry, which is to discover the causes of perceived phenomena. How so? Each scientist trusts that his system's behavior is caused and that the cause is analytically or experimentally discoverable. Yet the cause itself can be the object of investigation leading to a prior cause until we arrive at the conclusion (whether via general relativity, quantum mechanics, ecology, genetics, mathematics, or another scholarly discipline) that there is a First Cause upon which all phenomena and natural laws depend.[21]

Upon receiving the Medal of the American Institute of Chemists, Blum[22] said scientists have an obligation to benefit humanity materially, intellectually, and spiritually by making their findings accessible to the nonexpert public. He pointed out that because the study and application of science is spiritually uplifting, science and religion interface in man. Nonetheless each discipline has its own domain, with science describing *how* God works without being able to address why.

Concurrently, the president of the Sigma Xi research society used the words of Agassiz to describe in *Science* his joy of discovery over his first paleontological find:

Come wander with me
In regions yet untrod
And read what is still unread
In the manuscript of God.[23]

Meanwhile, on the very same page, the author lists several nontrivial research tips worthy of perusal today by even the best scientists.[24]

What was the scientific climate of those years, barely two generations ago? Scientists generally felt that the universe was totally accessible to analytical reasoning from known physical laws. Yet these same predecessors of ours were inclined to openly acknowledge the limits of scientific endeavor in the very journals which crowned their achievements. We, in contrast, know that a simple electron is untrackable, that protons themselves are vast composites of a variety of elusive, transient particles. In our day, science operates over some 40 spatial orders of magnitude[25] yet at none of these scales of resolution have we completely explained any system. Is it not odd that despite our acute awareness that the horizons of empirical knowledge are rapidly and continually receding,[26] some editors are today hard-pressed to allow an acknowledgment of the Architect who could engineer a cosmos which is at once orderly yet scientifically unfathomable?

Even within those constraints where one can say that a given system *is* precisely modeled, we still do not know why the discovered relationship exists. The existence of orderly patterns in nature is not itself an explanation; it begs one.[27]

Finally, modern advances in the sciences and statistics have been shedding light on the immanence of the Creator. For several centuries the monotheistic tradition has maintained a doctrine of continuous creation—that at every moment all aspects of the cosmos are brought into existence anew. While traditional scientists like Descartes recognized this in theory,[28] the recent discovery that subatomic particles are continuously and ubiquitously being created and annihilated exemplifies the process.[29]

Another important development is the discovery of analogous processes in physics and biology, suggesting a link which was long and actively sought after by Niels Bohr, whose life goal was the unity of all sciences.[30] In physics, Heisenberg's uncertainty principle allows large-scale phenomena to be largely predictable while events on the atomic scale are fundamentally indeterminate. This principle of order at large despite caprice in the small is demonstrable for nearly every scale and type of biological investigation.[31]

A particularly illustrative example is that the mental process of choosing to act is associated with a generalized electrical negativity across vast regions of the forebrain. This eventually causes an action but through no perceptible physical mechanism.[32] In this case, an act of physically unconstrained will imposes order on the neuromuscular system through an action which appears to be randomly generated. Analogously, physically unconstrained will could be infusing physical and biological systems, thereby imposing order through an apparently random pattern of small-scale events. While in the brain example, the will may be humanly generated, in physical and biological systems, the will may be Divine.

Ecological correlates of this phenomenon are environmental factors which track trends in distributed data. While the patterns are often predictable, individual loci are typically not, which is a sitution quite compatible with a continuous infusion of Divine Will into the system, directing the behavior of the system without disrupting the impression of randomness.[33] Thus randomness may be seen as an attribute of the system at a given perceptual level rather than its driving force, a scientifically unpalatable option on various grounds.[34]

The foregoing points are relevant to the Creator and science in general terms. The specific region of ecology also has its own connection to the Creator. Ecology is the scientific field which can be applied to conserve biotic populations, species, communities, and ecosystems. But ecologists cannot contribute to conservation unless conservation values are found in society. Ultimately, all reasons to conserve can be compromised and refuted except for the religious motive: Man has been made custodian over this global zoo by his Creator and only because of this do species have the right to exist.[35] This alludes to one of the Seven Laws of Noah which, when observed properly, give stability and durability to social as well as ecological systems.[36]

In summary, the validity of recognizing the Creator in scientific literature rests on several points: (1) there is a historical precedent extending to the present of recognizing the Creator in the mainstream of scientific endeavor; (2) it is a concept which unifies diverse branches of science; (3) it is consistent with all scientific findings; (4) a First Cause is basic to the universal scientific habit of seeking causal relationships; and (5) it is uniformly customary among scientific journals to allow authors to acknowledge those who made important, noteworthy contributions to the work.

From B'Or Ha'Torah 6E, 1987.

# 2

# The Religious Foundations of Science

*Avraham Kushelevsky*

It is commonly believed that natural sciences such as physics and chemistry, which deal with material phenomena, are based on objective observations of controlled experiments, while religion, which deals with spiritual phenomena, is entirely subjective and based on faith alone. It is no wonder, then, that any apparent contradictions between science and religion are inevitably resolved in rational debate in favor of science and its "proofs."

This attiude is very superficial. Science is no less a matter of faith than religion. Science assumes that there are internal associations between successive phenomena and, moreover, that the sequence of phenomena is significant. Science supposes that these associations are regular, constant over time, and discoverable through a series of observations during a short period. Similarly, science assumes that these associations have universal validity, obtainable not only in the laboratory but anywhere in the universe—the same in our galaxy as in a galaxy at the other end of the cosmos, the same in the recent and distant past as at any point in the infinite future.

Science posits that certain laws defined for masses of dimensionless points, lines with no width, and infinitely short time intervals embody objective truth about a real external world. Science posits that laws have a simple form and arbitrarily selects the simple explanation rather than the complex one if given a choice. For instance, although it is experimentally impossible to distinguish between a law operating according to $1/r^2$ or according to $1/r^{2}00000025$, the former formula is presumed correct because of its simplicity and aesthetic appeal. Without accepting these assumptions, the natural sciences as we know them today would be impossible.

These are reasonable assumptions, but they are not necessarily true a priori. Nor can they be experimentally verified, for the experimental method presupposes their truth. I am not criticizing these assumptions. I am only trying to emphasize that their acceptance is based on faith. It is not surprising that as little as a hundred years ago in England and Scotland the natural sciences were still called natural philosophy. He who says, "I believe only in what can be measured and do not rely on faith" is guilty of a fundamental contradiction in logic. He has nothing.

On the other hand, the religious scientist accepts these assumptions as an integral part of his religious worldview. He considers nature a manifestation of the wisdom of the Creator and science a means of drawing closer to Him. From the very first sight, nature shows differentiation, variety, and a multiplicity of phenomena. Science, which assumes simplicity and integration, actually brings man closer to God.

The religious person expects to find laws, logic, unity, and internal harmony in nature, reflecting the unity of the Creator and His wisdom. Thus, he believes in the existence of universal laws of motion that govern both the movement of the planets as well as the falling of apples on Earth, even before he begins to search for these laws by observation and experimentation. The unity of Creation leads to an acceptance of a deep inner uniformity in the structure of matter and in the existence of uniform field laws, even though his attempts to discover them over the last few decades have failed. He believes that there is nothing random in nature and that every phenomenon has a cause and a purpose.

For the believer, and especially for the religious Jew, faith in God and faith in science are complementary. But for the nonbelieving scientist, science is a riddle that has no solution.

From *B'Or Ha'Torah* 7E, 1991. This article first appeared in Hebrew in *B'Or Ha'Torah* 1. It was translated into English by Sam Friedman.

# 3

# The War between Religion and Science in the Nineteenth Century and the Change in the Twentieth Century

*Avraham Kushelevsky*

The war between religion and science reached its peak in the nineteenth century, a century characterized as "Godless" by Jacob Bronowski in his book *Between Religion and Science*. Albert Einstein wrote that in the nineteenth century "the basic unity between cultural, religious and secular institutions was lost and replaced by a senseless animosity." Previously, the attitude of science to religion had been one of respect and admiration. Isaac Newton wrote in the introduction to his *Principles*: "We can worship and admire the Creator of the Universe and this is the greatest outcome of the study of nature." Johannes Kepler saw his scientific endeavors as the service of God, and Blaise Pascal wrote of mystic revelations compared to which his scientific discoveries were of secondary importance.

15

What led to the change in the relations between science and religion, particularly in the nineteenth century? Most definitely, this change can be partly attributed to several geological and biological theories that became popular at that time and which seriously challenged the validity of the biblical description of Creation and the origins of life. Nevertheless, it is worthwhile examining the change in the relationship between science and religion in the light of several wider social developments occurring during that period. These developments were related to the war against the Church, the growth of the socialist movements, and the tremendous advances that took place in science and technology in the nineteenth century. A survey of these developments will allow us to understand further changes in the relations between science and religion in the twentieth century.

## THE WAR AGAINST THE CHURCH

In the eighteenth century, philosophers and thinkers waged a bitter struggle to free society from the yoke of the Church. The time for this was ripe. The Church had been corrupted and it had lost its moral authority. Its spiritual values had been perverted into fanaticism. The Church considered the body the source of all sin and tried to restore the spirit by repressing physical needs. Its medieval philosophy was out of place in a Europe that had experienced the Renaissance and was moving toward humanism. The Church itself degenerated into an institution that retained excessive material and political privileges and identified with the ruling strata of society. The war against the Church was presented as a war for progress and freedom, and one of the most effective weapons in the struggle was the ability of science to question and contradict religious dogma. Although only a relatively small group of enlightened individuals fought the war, this group won; their crowning achievements were the revocation of the excessive privileges of the Church and the separation of church and state in revolutionary France. The latter campaign spread from France throughout Europe, becoming a slogan for all antireactionary movements in the nineteenth century.

## THE TRANSFER OF EDUCATION FROM THE CHURCH TO STATE AUTHORITIES

An important outcome of the fight against the Church was the transfer of control over education from the Church, which had been responsible for all levels of education in Europe since the beginning of the Middle Ages, to the state authorities. It was partly a political move intended to undermine the Church. But it was also prompted by the rapid growth of industry, services, and governmental structures, which required a big expansion of the educational system. The Church, weakened by the reforms at the end of the eighteenth century, could not carry out this expansion.

In this way, the Church lost its most important tool for influencing Western society and the religious upbringing of youth. Instead, students at a very young age were being exposed to the antireligious ideas of their "enlightened" teachers in the state schools.

## THE FLOURISHING OF SCIENCE AND TECHNOLOGY

Concomitant with the political decline of religion there was an expansion of the impact of science on society. In the nineteenth century, science and technology soared to great heights. Classical physics reached full maturity with the discovery of electromagnetic field phenomena. In chemistry new elements were discovered and both the properties of compounds and the laws of chemical reactions were clarified. New fields such as bacteriology and immunology developed in biology, and such classical fields as physiology and anatomy were expanded beyond all recognition.

In the wake of scientific progress came accelerated technological development. The telegraph, the camera, the electric motor, electric lighting, and various kinds of internal combustion engines all enriched the life of the common man. Chemistry provided new pigments, medications, perfumes, and explosives. In medicine there was tremendous progress in the fight against infections and in the improvement of surgical techniques. In agriculture, new machines increased production a hundredfold. These achievements fostered an unquestioning faith in the power of science and created the illusion that in the near future technology and medicine would put an end to hunger and disease.

These developments changed man's view of the world. If, in principle, all natural phenomena could be scientifically explained, then there was no need for the assumption called "God." And since man could now rule nature, he need no longer turn to God for help.

## THE ROLE OF THE HUMANITIES IN THE BATTLE AGAINST RELIGION

The scientific method, so successful in such areas as physics and chemistry, which lend themselves to quantitative analysis, was then applied to fields less suited to quantitation—the study of the human mind, history, and society. The application of scientific methods to these realms gave rise to psychology and modern sociology. Although the scientific successes in these fields are still equivocal, it is tacitly assumed that by working systematically it is possible to make advances similar to those achieved in the exact sciences.

These successes have certainly done their share to change man's view of the world. Even the incomplete findings call into question such basic assumptions as the existence of a Divine soul, free will, God's intervention in history, and the Divine origin of the Five Books of Moses and the books of the Hebrew prophets. The scientific studies of these subjects suggest ways of developing a new morality based on "scientific" principles. Consequently, religion found itself facing a double onslaught by natural science and the humanities.

## THE EFFECTS OF POPULAR SCIENTIFIC LITERATURE

In the nineteenth century these new developments and challenges to the old order were brought to the attention of the general public by articulate scientists, who used their literary skills to popular-ize the acceptance of science and technology and the theories they implied. Men such as Aldous Huxley in England and Ernst Haeckel in Germany heralded the theory of evolution. In effect, their books spread antireligious venom under the guise of objective scientific re-porting. Their "scientific" disproofs of religion are superficial and at times even absurd. Haeckel, for instance, proves that there is no soul on the basis of the fact that it cannot be made liquid under conditions of low temperature and high pressure. However, with the extension of education to large sectors of the population, and especially to youth, even arguments such as these were found acceptable. Even Einstein was influenced by reading this popular literature and, at age twelve, became an atheist. Only years later, during middle age, did he come to recognize the existence of God.

## THE SOCIALIST MOVEMENTS AS THE SPEARHEAD IN THE WAR AGAINST RELIGION

It is possible that after the initial intoxication with the power and novelty of science had subsided, the "scientific" refutations of religion would have been resolved in a rational way; however, the big socialist movements of the nineteenth century now appeared on the scene. These movements postulated materialistic theories of nature in order to develop social theories. The idea that a supreme force intervenes in history and that one must surrender to it was not compatible with socialist theory. These movements saw any religion as an opiate of the masses that diverted attention from the real social problems. With all their revolutionary fervor and organizational skill, they renewed the war against religion. The war of ideas that had previously been waged by isolated individuals, scientists, and philosophers with no central direction was continued at the end of the nineteenth century by the socialist movements as a "holy war" to establish the rule of socialism throughout the world. In this war, the only legitimate religion was modern science.

## THE JEWS IN THE STRUGGLE
## BETWEEN SCIENCE AND RELIGION

Enlightened Jews who left the ghettos and worked for the emancipation of their people accepted the antireligious worldview that was prevalent among enlightened Gentiles. They also saw religion as a stumbling block to the integration of the Jewish people into European society, and they fought the religion of Israel as their non-Jewish friends fought the Church.

Some Jews became leading spokesmen of the socialist movements and, together with their Gentile comrades, they conducted the campaign against religion for the sake of world socialism. They saw all religions as divisive factors in human society, and they viewed the Jewish religion as a reactionary element hampering the establishment of a just, socialist, Jewish society. For many, the war against religion was really part of a rebellion against the authority of their parents' generation and an attempt to build a free self-image.

## THE CHANGES IN THE TWENTIETH CENTURY

In the struggle between science and religion, science was victorious. The English philosopher Bertrand Russell said as much when he wrote at the beginning of the twentieth century: "There was a prolonged confrontation between science and religion in which science proved its superiority." There were even predictions of the disappearance of religion.

The attack against religion started to wane because the fight no longer seemed necessary. However, just when religion seemed doomed, a surprising change occurred. Somehow religion survived and actually began regaining strength and credibility. The recovery of religion continues in our generation in the form of a renaissance of religious life on campuses and in the major science centers.

This change is partly explained by disappointment with the unfulfilled promise of the two major forces behind the antireligion conflict in the nineteenth century: socialism and science. The socialist movements simply have not lived up to expectations; the Communist party became more threatening and enslaving than the Church ever was. Paradoxically, the Church and religion are becoming the only unifying hope in restoring political freedom to the people of the Eastern Europe bloc.

## THE DECLINE OF SCIENCE

Science, too, has begun to lose some of its splendor. Hardly a single scientific theory that was considered basic in the nineteenth century remains accepted in its original form in the twentieth century. Quantum physics shifted the emphasis from deterministic to statistical theories. Einstein turned time into the fourth dimension and made the dimensions relative. The nineteenth-century theory of ether went the way of the eighteenth-century phlogiston theory. The atom, once believed to be indivisible, was found to be made up of subatomic particles. In the twentieth century, science is no longer certain that its theories embody absolute truth. Science has no guarantees that in another generation there will not be another transformation. Science is no longer infallible!

## SCIENCE BECOMES INCOMPREHENSIBLE

In addition, science has become less comprehensible. It is no longer the simple, clear statement based on test tube experiments. Particles are both waves and matter; matter is another form of energy; parallel lines meet in finite curved space. On the one hand, science has become mysterious and esoteric and puzzling, while on the other hand it has become impossible to spiritually identify with.

## A CRITIQUE OF THE FOUNDATIONS OF SCIENCE

In the twentieth century, science began a thorough examination of itself. It became evident that science had a logical limit and could never present itself as a system based on a few axioms and a limited number of operations. The uncertainty principle revealed that in the field of precise measurements there are theoretically insurmountable limitations. In the realm of scientific methodology it turns out that much of what was considered science is nothing but philosophy in scientific garb. Any theory that does not allow discussion of disconfirming phenomena cannot be called scientific. According to the English philosopher Karl Popper, this assumption deprives history and psychoanalysis of the right to be called science. The "scientific" theories of evolution and cosmology, which are based on unverifiable extrapolation, are similarly afflicted. Gods that aren't sure of themselves slowly rot and fall; they don't attract worshippers anymore.

## THE NEGATIVE EFFECTS OF TECHNOLOGY

In the twentieth century the negative effects of technology have become apparent. True, technology has succeeded in landing a man on the moon, but it has not solved the problem of air pollution in the big cities. Technology has afforded man a terrible potential for self-destruction. Poisonous gases, germs, and atomic bombs are ready to wipe man off the face of the Earth; factories have polluted the environment in a way that imperils the climatic equilibrium of our planet. Pesticides are disrupting the natural ecological balance and threaten to end the spring symphony; visions of horror arise from the test tubes of genetic engineering; the centralization of information in computer data banks points the way to an Orwellian 1984-type controlled society. Man fears the monster that technology can become.

It seems that the real problems confronting the world are precisely those that deal with morality and the soul—those that have always confronted us. Mankind's existential dilemmas have yet to be solved. The problems of good and evil, and life and death, are as pressing now as they were two thousand years ago. Man's existence today is dependent on his ability to overcome his drive toward self-destruction and the murderous competition between societies. Man's happiness is dependent on his finding spiritual ideals that give his life purpose and content, for man cannot live on bread alone. In these matters, science and technology offer no answers; but religion does.

## CHANGES IN THE JEWISH WORLD

In the twentieth century there has been a social change in the Jewish world. Before World War II, Jewish youth chose either the Torah (the *yeshivah*) or science (the university); few individuals lived in both worlds. After the war, especially in the United States, a stream of *yeshivah* graduates started to flow into the universities to study science and engineering. This was a practical reaction to the increasing economic demand for scientists and technologists, as well as to the social status and income that these professions offered. Gradually the general public is becoming aware of learned, observant Jewish scientists occupying high-ranking positions in scientific institutions and centers of power. The religious Jew is no longer stereotyped as a do-nothing, divorced from the world. No longer is the question heard, "You are a scientist—how can you be religious?" Religious scientists are dealing with the problems that religion presents to science and are finding both old and new answers that appeal to their colleagues.

## CONCLUSION

The conflict between science and religion became acute in the nineteenth century. The war against a corrupt Church, the ascent of science and technology, and the false promise of socialism led to the rejection and weakening of religion.

In the twentieth century, a balance has been restored. We know that technology exacts a price for its service to mankind; science offers no real answers to the ultimate questions; and freedom is incompatible with enslavement to a political-economic system that denies God.

From *B'Or Ha'Torah* 7E, 1991. This article first appeared in Hebrew in *B'Or Ha'Torah* 2. It was translated into English by Sam Friedman.

# 4
# The Torah–Science Debates: Some Random Thoughts

## Velvl Greene

### THE DEBATE CONTINUES

Most of the participants in the science–religion debate think that it is only about 150 years old and that it originated in the Europe of Darwin and Wellhausen. Indeed, much of the debate—as we identify it today—does contain elements that were newly introduced during the Darwinian era, such as biological evolution and cultural anthropology. But the basic disagreement predates, by far, this recent past. The science–religion controversy is a venerable and long chapter in the drama of world and Jewish history.

Galileo's persecution by the Inquisition of his church occurred in the early seventeenth century, but Maimonides' efforts to reconcile Aristotelian science with Torah were recorded five hundred years earlier. The debates described in the Talmud between the rabbis and the "heretics" (who depended on their own empirical observations rather than revelation and tradition) date back to the very early years of the common era, some eighteen hundred years ago. It does not require too great a flight of the imagination to visualize Moses' appearance in Pharaoh's court as an early science–religion debate: one side accepting the validity of tradition, revelation, and miracles, with the other side trusting only logic, reason, and empirical "proof."

For some reason, the centuries-old debate still continues. In Jewish life, it seems to continue with unabated intensity. For example, the mediocre Anglo-Jewish newspaper that tries to serve the social and informational needs of my small Diaspora community devotes an inordinate amount of front-page space to science–religion "news." Sometimes an article on science appears by one of our more liberal rabbis trying to establish his credentials as a "modern" philosopher; at other times, the paper reports on the dangers that accompany the efforts of "creationists" in America and Israel to challenge contemporary science curricula. Howsoever presented, the controversy is still viable. Attempts to defuse the issue and to resolve the disagreements in books, journals, and conferences have not really succeeded. The Torah–science debate, it seems, is with us today as much as it was a century ago, or many centuries ago.

## THE DEBATE CONTINUES—BUT THE DEBATERS CHANGE

This is not to say that the same banal arguments are repeated century after century and generation after generation. Both the actors and the script have changed. The arguments in ancient days pitted revelation against reason; religion and Torah were regarded as dogma while science was labeled enlightenment. One side represented conservative, reactionary authority, while the other side claimed to speak for liberalism, empiricism, and freedom. Today the debate actually deals with very different subjects. The labels may be the same (depending on which side uses them), but the arguments really reduce to the validity of the scientific method and the true meaning of the Torah. Much of the controversy is based on interpretation and misinterpretation of natural phenomena and probability and the relevance (even the feasibility) of Divine revelation. Paradoxically, the "science arguments" have become more and more dogmatic and the "science spokesmen" have become more and more authoritarian, while the "Torah spokespersons" become the challengers of conventional wisdom and the advocates of free inquiry.

Perhaps the arguments strayed from the original because the participants in the argument have been replaced. Folklore would have it that the proponents of science are scientists: physicists, chemists, biologists, and engineers with lab coats and complicated equipment. Arranged against them—again according to folk wisdom—are the bearded, black-hatted rabbis whom the *Jerusalem Post* labels "ultra-Orthodox" and whom most of us consider to be Torah scholars if only because of their dress.

It may have been like this at some time in the past. It is not today. The debate might be going on in the *Jerusalem Post* and in the Anglo-Jewish diaspora press, but it is not heard in the science lab where I work.

## PERSONAL CONFESSION

To tell the truth, the Torah–science debate never played a serious role—either positively or negatively—in my own professional career or in my slowly developing "observant" life-style. The debate left me cold thirty-five years ago, before I kept *Shabbat*, and it isn't a priority item in my life now that I do.

When my scientist colleagues evaluate—or criticize—my research, or writing or teaching, they completely ignore my association with Lubavitch. It appears that my religious outlook is of greater concern to the editors of the *Jerusalem Post* than to the editors of *Science*. (Maybe this is a defect on my part. Maybe there should be a more evident influence of Habad philosophy on my professional work. Perhaps it should modify my research and alter my capacity to reason.)

But really, there are only two kinds of science: "good science" and "poor science." I've always tried to do "good science," before, during, and after my stages of indoctrination into Torah and the *mitzvot* (commandments). Sometimes I succeeded, sometimes not. But my honest belief in the literal truth of every word in the Torah hasn't significantly affected the success/failure ratio of "getting funded" or "getting published." If there is a science–religion conflict, I missed it in both my scientific career and my Torah career.

Moreover, I don't know of any good scientific establishments—in academia, industry, or government—where the Torah–science debates are any more relevant than they were in my life—certainly not as relevant as implied by the newspapers of my Jewish community. In the science world I know the quality of data and the quality of reasoning used to interpret these data are the criteria of acceptability, not one's diet or head covering. Most of the good scientists I know, like most of the people I know, are as yet nonobservant in their personal lives. (Though I am often impressed by the number who are!) However, their pro-Torah or anti-Torah biases, if such exist, don't really shine forth. Such biases are possible, of course, but good scientists don't participate in the debate. Instead they let rabbis take the "side of science."

As I search my memory, I do recall an incident where the science–religion controversy had more than a passing impact on my work. In a class of 120 nursing students, one young lady insisted that the proper "community health" approach to infectious disease should be prayer. On her examination paper she explained clearly that epidemics were Divine punishments visited on communities because of their sins. It is a testable hypothesis and should be studied by epidemiologists. Indeed, it might be the proper answer in a theology course. But it was the wrong answer in my course in public health. She failed the examination. She appealed on the grounds of my antireligious bias. She lost the appeal.

It may be significant that the controversy is ignored not only in the laboratory where I work but also in the *minyan* where I pray. This may be the real paradox. One should expect the issue to explode here or at least to generate acrimony. Consider the scenario: twenty to thirty veteran Israelis—most of them academics, some even world-class scientists, some with rabbinical ordination from rigorous *yeshivot* and a few who came along later in life as I did—all praying or learning together or singing *zemirot* of *Shabbat*. Most come across as pretty sincere in their belief, their observance, their professions. But the classical conflicts enumerated above don't seem to be very important. The professor of statistics is not visibly disturbed by the probability of miracles. The researcher in anatomy uses the *Shulhan Arukh* (Code of Jewish Law) rather than his textbooks to determine the kosher status of a fowl. The chemist is more concerned with the proper pronunciation of the Torah reading than with the ratio of carbon 14 to carbon 12 in ancient ax handles.

There are many who are bothered by the paradox I just described. If real scientists and observant Jews aren't debating the Torah–science issues, then who is? If we live in a world where genuine scientists pray and where sincere Torah observers are elected to national academies, then who is keeping the debate alive and why? I submit that several major constituencies in Jewish life have (or once had) a vested interest in the debate. I submit further that their motives for participating in and perpetuating the debate deal less with a search for truth than they do with self-preservation. And I submit, above all, that the biases introduced by these movements have so obfuscated the basic controversy that the average Jewish layman doesn't even know what the argument is really about.

## THE FIRST CONSTITUENCY: THE IMMIGRANT GENERATION

One constituency was my parents and their generation who grew up in the last decade of the nineteenth century. They made up their minds about eighty years ago, and nothing discovered in science or elucidated by Torah scholars since then modified their conclusions.

To my parents, science and technology were not mortal enemies of Torah Judaism; they were its natural successors. In their eyes, science and technology represented progress, promise, and the New World. They didn't really understand what Liste, Kelvin, Darwin, and Freud were saying, but they believed in those men. On the other hand, they didn't really understand what the Torah was saying either. But they associated the Torah with the small Eastern European Jewish town, mud, superstition, and restriction. The generation that opted for Zionism and socialism, for culture and freedom and liberalism, replaced the superstitions of the Old World with the superstitions of the New World. They didn't do it intentionally, but they did it nonetheless: they deified science and technology and built altars to these new gods. In some cases they sacrificed their children on these altars. They certainly sacrificed their heritage.

This is not the place to analyze the causes and dynamics of that sociocultural revolution. But I'm sure of two things: the reasons they gave for discarding Torah (they would prefer to use the word *updating* because *discarding* has such a pejorative ring) had precious little to do with Torah; and the reasons they gave for deifying science did not derive from knowledge of science. The point to be made, however, is that it happened. The support of science, the blind acceptance of the validity of scientific statements, and the equation of science with truth (at least science and truth as they were perceived at the turn of the century) became an important thread in the fabric of modern Western Judaism.

The situation is really a little pathetic to one who follows historical developments in science. The Jewish community has changed profoundly since it emerged from the *shtetl* mud one hundred years ago. The twentieth century, which experienced both Holocaust and rebirth, has irrevocably remolded Jewish values and perceptions. Zionism, socialism, and communism are no longer the major issues of our existence, even in Israel. Dreams about internationalism and the United Nations and equality among the "family of nations" grow more sour every time the General Assembly meets. The promise of America and its political-economic mirage are being critically reexamined in the light of affirmative action and Jesse Jackson. But Jewish attitudes to science—especially as they conflict with Torah—are quite unchanged. They remain fixed—squarely in the last decade of the nineteenth century.

## THE SECOND CONSTITUENCY: THE MODERN RELIGIOUS ESTABLISHMENT

The "modern" Jewish approaches to Judaism in America also draw some sustenance from the Torah–science conflict. However, unlike the immigrant generation, they do not consider science to be the successor of Torah; instead, they try to rewrite the Torah in order to show its fundamental agreement with modern truth—science. In this way, they can justify two of their basic claims to legitimacy on the Jewish scene, as guardians of religion, Torah, and tradition on the one hand and as modern replacements for the Neanderthals who teach that Torah is eternal and immutable on the other.

Interestingly, all of the modern Jewish outlooks, regardless of their theological disagreements, have an almost identical approach to the challenge of science to Torah. According to this approach, the Torah writes allegorically whereas science proves things; the Torah was designed to teach morality whereas science teaches facts. As scientific discoveries are made, the Torah must be reinterpreted to avoid conflict with new discoveries. Jews are no longer obligated to accept Torah narratives in a literal sense, says the liberal theologian. Instead, they are ready to accept any new theory, any new statement, any new version of reality—as long as it has the imprimatur of science. Thus, Torah must be preserved for its moral and humanistic and even literary value, but its primitive elements and nonscientific pronouncements are dated and disposable and modifiable. The bottom line, actually, is to deny completely that the Torah is capable of saying anything it really intends to say and to grant credibility only to what is accepted by the sophisticated, liberal establishment of Cincinnati, New York, and Los Angeles in the eighties.

In liberal theology there are very few absolutes. Compromise is a virtue, and tinkering with tradition is a way of life. It should not be any surprise, therefore, to recognize that in the Torah–science conflict, modern religion doesn't even pause to challenge science. There is no claim made by a psychiatrist with credentials that is too bizarre, no hypothesis by an astrophysicist that is too preposterous. All science and all research are accepted as valid sources of truth and the Torah is stretched and twisted on the Procrustean bed of compromise to conform to the new discovery.

Even better is any new discovery which conflicts with the "inconvenient" laws of Torah. The hypothesis that dietary laws were really ancient public health regulations, for example, killed two birds with one stone. It showed that the Torah was once valid (even advanced!) and it simultaneously permitted us to replace the old laws with modern sanitary codes formulated by the high priests of the Federal Food and Drug Administration. (No argument against this hypothesis, even from recognized experts in public health and medicine, seems to be effective. Science —even poor science—wins against anachronisms every time. Dietary laws can be accepted and observed as a noble tradition or as a historical link. But as a Divine decree it loses any real clout.) This is also true with the Torah sex laws. Since it is known that all ancient people had sex taboos, the Torah can just be considered another historic document. Therefore, some parts can be accepted (for hygienic or moral reasons) and others can be rejected. The concept of Commandment was the first casualty of the rewriting of Torah—long before the concept of the age of the Earth.

The question can be fairly asked: Why modify the Torah? What is the benefit of rewriting it? Why not discard it altogether like my parents' generation did? Indeed, there are those, to the left of the Jewish American religious establishment, who claim that any observance of Torah—moral as well as ritual—is anachronistic. Beyond the liberal rabbi is someone more liberal who questions the need for a rabbi altogether. That is exactly why the liberal religious still need the Torah or whatever emasculated parts they have chosen to retain. The Torah–science controversy has become a Torah–science "synthesis"—the perfect ecological niche for those among us who "want to do their own thing" and simultaneously remain traditional. We want to retain some kind of Jewish institution, a synagogue or temple, if only because the Gentiles have their churches and look with suspicion on the unchurched. Or we want to salve our conscience after visiting the Yad

Vashem Holocaust memorial, or to justify paying the salary of a rabbi. We will use whatever Torah is needed to maintain identity and will deny the remainder because of scientific discovery. Very rarely does the formula include such components as the Will of God—the desire and intent of the One Who gave us the Torah in the first place.

The preceding model can be used to describe the Torah–science conflict in Israel today. The compromises and selective utilization of Torah by those who manage Israeli culture, politics, and education are much more exciting than the dull American controversy. But I won't touch it until I take up permanent residence in Israel and can speak as a participant rather than a spectator.

## THE THIRD CONSTITUENCY: SOVIET RUSSIA

There is yet another milieu in which the Torah–science debate is taken seriously: in Soviet Russia and other totalitarian states where the value of the individual and his spiritual component is basically denied. Here, of course, the debate is not left to rabbis who are untrained in science and who only want to compromise a little bit. Here, the fight is conducted seriously, urgently, and systematically with the whole power of the state mounted on the side of "science." And the fight has been remarkably successful except for a small handful of stubborn men and women who are the true holy people of our generation. But in Russia the fight is understandable. Torah is the antithetical paradigm of the Soviet philosophy. Torah and dialectical materialism are in absolute and uncompromising opposition to each other at every possible point along their interface. That is why the Soviet government deems it imperative to eradicate Torah, its values and its narratives, its heroes and its ideals. For communist ideology to win, it must not only eradicate the Ten Commandments; it must also destroy, completely, the idea that there is a Creator and that there was a Creation, that there was a beginning and that there is one God and many worlds. They are not worried about being considered "old fashioned." The very ideas go to the heart of an attack on the "Soviet premise" and its continuity. Thus the Soviet government, for its very survival, uses science twice: once as a weapon to contradict Torah, and once—as the model of neutral values—to replace Torah.

## THE ISSUES AND THE BAGGAGE

It appears from the preceding, that the Torah–science conflict— or what is labeled as the Torah–science conflict—is not really new, is not of much concern to either scientists or Torah scholars, and is obfuscated by a great deal of extraneous baggage. The participants in the debate are not really motivated by a search for the truth. Instead, there are those who want to replace Torah with science, others who want to rewrite or reinterpret Torah to make it conform with science, and still others who use science to destroy the essential validity of Torah. And superimposed upon this is a lot of anger and basic ignorance. It is remarkable how few of the "pro-science" arguments are based on valid, reproducible, scientific data. Scientific hypotheses and speculations galore are cited, often by spokespersons whose deepest involvement with science is the National Geographic or a college textbook. But first-class thinkers and philosophers of science—who appreciate probability and uncertainty and the limitations of scientific experiments— rarely participate in this debate. By the same token, the "pro-Torah" arguments are too frequently out of context quotations from an English translation of the Bible. The commentaries of one hundred generations of Torah sages are almost universally ignored. We need go no further than the commentary of Rashi on the very first sentence in the first chapter of Genesis. How much acrimony would be spared, how much wasted debate about big bangs would be avoided, if Rashi's explanation (that the Torah starts with the story of Creation not to provide a cosmological explanation of mechanisms but rather just to establish the proprietorship of the world and the right of the Proprietor to allocate portions according to His Will) were taken at face value! Why is it easier to accept the authority of Lord Balfour to justify Jewish occupation of Eretz Yisrael? Perhaps because Balfour was English? More probably because Balfour's declaration doesn't imply acceptance of other things. Implied by Rashi is a Creator Who is concerned, Who communicates His instructions by means of a Torah, Who makes certain demands.

Thus the essence of the Torah–science debate is the personal and communal consequences and obligations that are incumbent on every Jew, on every human, if indeed the Torah is true. The debate ceases to be an exercise in philosophical speculation when one's very fate depends on the outcome.

It isn't really relevant whether the world was created in seven days or seven thousand years or seven billion years. It might be fun, but it isn't important to win a debate about Big Bang fifteen billion years ago or ten Divine statements 5,745 years ago. And it is exciting but essentially nonproductive to argue about the validity of whatever evolutionary theory is in vogue today. The basic question is the same one today faced by Moses and the Sages of the *Mishnah* and Maimonides and the Rebbe of Lubavitch in that communist prison in 1827. The question they faced was that of the ultimate source of truth and the ultimate guide to human behavior. Is it human reason, observation, and conjecture or is it Divine instruction as revealed in the Torah?

The Torah–science debate consequently becomes an urgent issue when it leads people away from Torah. If a Jew becomes convinced that the discovery of dinosaur bones is an excuse for not wearing *tefillin* (phylacteries), this is much, much more serious a problem than scoring points in a debate. And the Torah–science debate, or what is labeled as such, is important if such arguments as those dealing with the age of the universe distract or delay someone from accepting fully the *mitzvot* which the Torah obligates him or her to do. Science and technology have so much to offer mankind; what a terrible waste that misconceptions and mistakes serve as a crutch for the nonobservant and as an obstacle to those who want to return to the Torah.

This latter point might need reemphasizing and will be treated more fully in a later essay. It isn't the basic disagreement between good science and Torah that leads to acrimony and to transgression of Torah. It is completely consistent with Divine Providence if a scientific observation or experiment reveals a previously hidden truth about our physical universe or our health. There can be no inconsistency between truth and truth, reality and reality. On the other hand, the mistakes in science (or the misunderstandings of Torah) are the stumbling blocks in this dialogue. The self-serving hypotheses that masquerade as science and the "scientific" dogmas that have been generated over the last century cloud an honest examination of the issues. They inhibit the honest searcher for truth from exercising his free will.

There is a great need to remove some of this baggage. There is an urgent requirement to clear the tangled underbrush from the landscape. We can have an honest debate on the issues and controversies in the Torah–science arena only when we have neutralized the biases, examined some of the key dogmas, and clearly articulated the arguments.

If we do this, we might discover that there is no debate, no disagreement, no need to compromise. We might discover that, after all, science is not a crutch for the nonobservant and is not an obstacle to those who have not yet returned.

From *B'Or Ha'Torah* 6E, 1987.

# 5

# What We Cannot Know

## Paul Rosenbloom

Eddington[1] gave a striking illustration of the limitations of knowledge. He tells of a scientist who drops nets all over the seven seas, examines his catches, and announces, "There are no fish less than an inch long." An observer examines the nets and replies, "Since the mesh of your net is one inch, you couldn't possibly catch any smaller fish, even if there were any." This shows how, by analyzing the methods of investigation, one can prove that certain kinds of things are not knowable. Eddington points out that the deepest and most certain principles of science are all statements that certain things are impossible to know.

People whose knowledge of science comes from popularizations and elementary courses usually have a quite contrary impression. They "know" that "science has proved" many things. They believe that every problem can and ultimately will be solved by reason and observation as more powerful and refined methods are developed, that our ignorance of certain things is only temporary, and that everything is in principle knowable. But the argument of Eddington's parable shows that there are intrinsic limitations to what can be known.

37

Thus, in order to find out what *is* accessible to science, we must examine scientific method. Many textbooks[2] outline the scientific method as:

1. Observation
2. Empirical laws
3. Theory
4. Prediction
5. Test by further observation
6. Revision of theory

As many historians of science have remarked, this paradigm is over-simplified and rarely fits what has actually happened. Still it may serve as a starting point of our discussion.

## A THEORY MAY BE DISPROVED
## BUT NEVER PROVED

The most important consequence is that a theory may be *disproved* by the failure of its predictions to fit the facts, but it can never be *proved*. Predictions which are confirmed make a theory more plausible, but there may be other theories which also explain the observed phenomena. Even Newtonian mechanics, which was very successful for more than two hundred years, was finally rejected in favor of relativity.

Basically, the only facts that we really "know" are what we observe with our sense organs. I will not discuss here such problems as: can we prove that there is an external world (Berkeley, Russell), can we know whether we are dreaming or awake (Bridgeman), do our sense perceptions have the same structure as the "real world."[3] However, I do wish to point out several ways in which scientific observation is more complicated than is indicated by the above simple sketch of scientific method.

First, all observations of the very small or very large require extensions of our senses by means of instruments. So we always need a preliminary theory of our apparatus in order to interpret the observations.

Thus, when we determine the position of a planet, we assume that light travels in straight lines in a space of some assumed geometry, and we make use of the optical properties of the lenses and mirrors in our telescopes. Until achromatic glass was developed about 180 years ago, microscopists reported many fantastic phenomena which we now consider to be artifacts of their instruments.[4] Watson and Crick did not "observe" the double helix structure of DNA. They made a Fourier analysis of the X-ray diffraction patterns as recorded on photographic plates and found that the double helix hypothesis fitted the data very well. Millikan did not measure the charge of an electron. He calculated the charges on oil drops from their motion in an electromagnetic field and found that they were approximately whole number multiples of some fixed charge (after discarding the data which did not fit!).

The example of DNA, just cited, illustrates how the gap between the observation and the interpretation has increased with the development of science. In elementary particle physics one actually observes streaks on photographic plates. We interpret them as evidence of particles being created or annihilated. The underlying quantum electrodynamics is a highly abstract theory, many of whose constructs cannot be observed even in thought experiments.

## IS THERE AN EXTERNAL WORLD INDEPENDENT OF THE OBSERVER?

In a little book, *What Is Life?* Schrödinger[5] discusses the contrast between the worldview of classical science, going back to the Greeks, and that of modern science. According to the former, there is an external world independent of the observer. While we accept the inevitable errors of observation, these can, in principle, be made arbitrarily small by the eventual improvement in our instruments. The ideal is a description of phenomena independent of the observer.

In modern science, we recognize that an observation results from an interaction between the observer (and his instruments) and the system observed. In the process, the states of both change. Heisenberg[6] shows, by analysis of these changes, that there are intrinsic limits to the accuracy of measurements. For example, it is impossible to measure both the position and the momentum of an electron at a given time to within arbitrarily small errors.

Schrödinger, in the cited book, examines the problems: If observations depend essentially on the observer, how can there be objective knowledge? Why do observers' descriptions of phenomena agree so well? After a detour through Hindu philosophy he arrives at the explanation that our minds are parts of the Divine Mind. He is the only true observer, and what is common to our observations comes from Him. The concept of the Divine Mind, as presented by Schrödinger, is very similar to the hasidic concept of the divine soul.[7]

We alluded above to the step of formulating empirical laws, which generalize our observations, for instance, the derivation of Kepler's laws of planetary motion from Tycho Brache's data. This always involves going beyond the facts, either by interpolation or extrapolation. The latter is especially treacherous. The further we depart from the known range, the less reliable is the extrapolation. Thus, at very low temperatures, properties such as electrical conductivity are completely different from what they are in the ordinary range; high temperature chemistry is radically different from ordinary chemistry. Newtonian mechanics broke down at very high velocities and subatomic distances, leading to relativity and quantum mechanics, respectively.

## "PREDICTION" OF THE PAST

Extrapolation backwards in time involves special difficulties. In such phenomena as heat conduction, diffusion, and processes affected by chance, "prediction" of the past from the present is a "badly set" problem.[8] For instance, there are infinitely numerous and very different temperature distributions on a thin rod which lead, an hour later, to distributions differing by an arbitrarily small experimental error. Hence, Lord Kelvin's estimate of the age of the Earth from its rate of cooling and present temperature is completely unreliable—besides the fact that he did not know the heating effect of radioactivity. In Newtonian mechanics, if the present state of the solar system as well as all the forces acting on it are known exactly, then we can predict backwards as well as forward for any finite time interval. But consider the very simple system of a ball rolling frictionlessly on a rectangular billiard table with perfectly elastic walls. If we know the dimensions of the table and the present position and velocity of the ball only to within some experimental error, then the ball could have been almost anywhere on the table an hour ago.

## ORIGIN OF THE UNIVERSE—MUTUALLY INCOMPATIBLE THEORIES

So far we have considered some general limitations of the scientific method. Let us now turn to two of the particular problems which arise in the discussions of religion and science: the origin of the universe and evolution.

Until recently, the most widely accepted geological theory has been the uniformitarian principle, which explains the past by using laws which are known to hold now, assuming that they have acted in a gradual, uniform manner throughout the past. Of course, the obvious objection is that all scientists agree that conditions of temperature and radioactivity were totally different at the time of the formation of the earth. Indeed, lately the rival catastrophe theory, which explains the past in terms of fairly frequent large sudden changes, has been revived by Gould[9] and other leading researchers.

At a meeting organized by the Royal Society of London and the Royal Astronomical Society in 1972,[10] it seemed that all the researchers had mutually incompatible theories on the origin of the solar system. For a survey of the many theories of the origin of the earth, we may refer to Hoyle's address.[11] There are three main types of theory of the origin of the moon: capture of the Earth; breaking off from the Earth; and simultaneous formation of the two. Their proponents all find support in the analysis of lunar rocks.

Until comparatively recently, Aristotle's principle that matter can be neither created nor destroyed was generally accepted and was a major objection to the Torah account of the creation of the universe out of nothing. But in quantum field theory we now consider that creation and annihilation of elementary particles occurs everywhere and at all times. The most widely accepted theory of the origin of the universe now is Gamow's Big Bang theory,[12] according to which all matter in the universe was created suddenly out of nothing within about half an hour. The main problem with the Torah account remains the question of the date. Jastrow[13] has pointed out the almost religious discomfort of many scientists with this theory because it seems to raise the question of "Whodunit?" Incidentally, this theory is very reminiscent, qualitatively, of the description of the creation of the universe in the *Zohar* and Luria.[14]

Regarding the history of life, a distinguished biologist, D. D. Davis,[15] wrote "the facts of paleontology conform equally well with other interpretations, e.g., divine creation, innate developmental processes, Lamarckism, etc., and can neither prove nor refute such ideas."

## THE RABBIT-WARREN OF EVOLUTIONARY JUST SO STORIES

There is no evidence that evolution has actually occurred. Eldredge and Cracraft[16] discuss "the rabbit-warren of intestable storytelling (scenarios) which comprise much of past and present evolutionary theory" and call them "just so" stories. In fact, according to the dominant neo-Darwinist school, which holds that evolution proceeds by accumulation of small changes, there must have been, for example, animals intermediate between reptiles and birds. They explain the absence of evidence in the fossil record by its essential incompleteness and the rarity of these intermediate creatures. The saltationists (e.g., Løvtrup),[17] advocating evolution by sudden large changes, say that such evolution at the class level occurs only once in fifty million years, so it would be a miracle to observe it. Thus, neither hypothesis is testable by evidence.

Many generations of such rapidly reproducing organisms as fruit flies and bread molds have been bred in genetic laboratories since 1910 and even subjected to mutation-producing agents since 1927. Yet no mutation to a different species has ever been reported.

Viable and fertile offspring with different numbers of chromosomes from the parents are so rare that their occurrence is close to a miracle.[18] (I am ignoring the phenomenon of haploidy and diploidy which do not raise fundamental questions.) How many such events must have occurred to produce the millions of species of insects with widely varying chromosome numbers, from some hypothetical common ancestor?

Evolution, at present, is not a scientific theory but rather a conceptual scheme for interpreting the fossil record and the similarities and diversities among organisms. Thus, any observations whatsoever can be fitted into the scheme, and it would be impossible to conceive of evidence contradicting it. The few valiant attempts to make it precise enough to be testable (e.g., Wright,[19] Løvtrup[20]) still leave big gaps between theory and observable facts. The vesicular eye (with retina, pigment layer, lens, and cornea) has been a classical challenge to evolutionists. Rensch[21] has devised the most plausible scenario I have seen to account for it. But he completely ignores such questions as the number of gene changes involved, the probabilities of such a sequence of events as he envisages, and the expected time for such a development. Furthermore, he himself admits that this occurred in such diverse groups as coelenterates, annelids, echinoderms, onychophores, gastropods, cephalopods, and vertebrates. To the question "Could it have happened?" his sketch gives an answer: "Maybe, given enough luck and time." To the questions "Did it happen?" and "How could we find out?" it is completely silent.

## BELIEF IN A THEORY IS AN ACT OF FAITH

A theory is a working hypothesis. If predictions agree with the facts, the theory becomes more plausible but never certain. This agrees with the views of all scientists and most philosophers, e.g., Popper, Waddington, Eddington. Thus, belief in a theory is an act of faith, not a matter of reason and observation.

Many attempts have been made to reconcile Torah with various scientific theories. There has even been a new translation based on *midrashim* consistent with the Nebular Hypothesis. But we can learn a valuable lesson from Maimonides.[22] He was writing for the Jewish intellectuals of his time who had learned modern science (Aristotle) at Moslem universities and had been shaken in their faith. He devoted much effort to explaining how an enlightened person could still accept *Creatio ex Nihilo*. If he could only have waited for the Big Bang theory, he could have spared himself the trouble. Any reconciliation of Torah with a particular theory is dangerous. A leading cosmologist, E. A. Milne, remarked that a book on physical theory is now usually out of date in ten years. In science, "the only things that are permanent are the experimental results."[23]

Scientific laws are all laws of transition. They say that if a system is in a certain state, then it will be in such-and-such a state at a nearby later time. Backwards prediction is a deduction that if a system existed at a certain time in a certain state, and these laws held in the intervening time interval, then it should be in this state at present. We can neither know when the system came into being nor whether these laws held throughout that period. This is akin to other laws maintaining, in effect, that certain things, like the consistency of arithmetic,[24] and the position and momentum of an electron at a given moment, are unknowable. (This is related to an argument attributed to Russell.[25] Incidentally, for a theory of natural laws varying with time, see Dirac.)[26]

For example, suppose I handed my clock to an expert in mechanics, let him examine it thoroughly, and then asked him, "What was the position of the hand twelve hours ago?" He might answer, "If it was wound up and in good working order during that time, then the hands were in the same position as now. But I can't tell when it was wound up!"

Rashi's commentary on the sixth day of Creation suggests the possibility that God created the entire world during the first six days and put everything in its proper place. He then set it all in motion when He made Adam. The hypothesis that God created the universe complete with fossils, rocks, etc., as they should have been 5,743 years ago is consistent with all scientific laws. The choice between this hypothesis and one like the Big Bang theory is a matter of faith, not of scientific fact.

## BELIEF IN THE TORAH DOES NOT LIMIT
## OUR FREEDOM TO INVESTIGATE
## THE LAWS OF NATURE

In the first benediction before the *Shema* in both the morning and evening services, we thank the Lord for creating an ordered universe. In fulfilling the injunction to "subdue" the earth (Genesis 1:28), we may explore those aspects of His wisdom which are accessible to us. Thus, belief in the Torah does not limit our freedom to investigate the laws of nature.

Someone may object, "Why should He begin the Creation with a highly developed universe, when He could just as well have begun with nothing and worked His way up through quarks and amoebas to that state?" We might turn the question around, "Why should He wait four billion years when He has the power to start with the state in which He was interested?" This objection is just as impertinent as asking why Shakespeare began with Hamlet instead of prehistoric Denmark. If the time when He made the world is inaccessible to science, how much more so are those purposes which He has not revealed to us?

From *B'Or Ha'Torah* 3, 1983.

# II

## Specific Applications of Looking at Science and Art Disciplines in the Light of Torah

# Physics

# 6

# The Creator in Creation

*Naftali J. Berg*

## FROM DETERMINISM TO INDETERMINACY

According to the mechanistic and deterministic physics of the nineteenth century, all, or almost all, causes and effects were known. Given an absolute reference framework plus completely deterministic mechanics, the entire future was considered predictable in mechanical terms. This viewpoint did not exclude the Creator, but it did exclude the concept of Divine Providence. The Creator was considered necessary only for having started the clockwork of creation. Twentieth-century physics changed this viewpoint. The two areas of change which we shall discuss here are that of quantum mechanics in the microcosmic domain and that of general relativity in the macrocosmic domain. Let us first consider quantum mechanics. We have the Schrödinger equation:

$$-\hbar^2/2m \; \bar{\nabla}^2\psi + V\psi = i\hbar \; \delta\psi/\delta t \qquad (1)$$

where the wave function ψ is interpreted as a not directly observable quantity which relates to the probability of finding or locating the "particle."

This is a linear equation and solutions can be superimposed. Associated with this equation is Heisenberg's uncertainty principle, which states that the location $x$ and momentum $p$ of a particle cannot simultaneously be specified together to accuracy better than $\hbar$ (Planck's constant). In mathematical terms:

$$(\Delta x)\ (\Delta p) \geq \hbar \qquad\qquad (2)$$

In other words, an uncertainty—or better yet, an indeterminacy—entered physics. Certainty was replaced by probability.

From a religious point of view, this was a step in the right direction. But this was an *extreme* step, as the eventual interpretation of the uncertainty principle was not only that there was uncertainty in measurement but also that there was a *basic* uncertainty in nature expressed in Heisenberg's philosophical writings. This uncertainty principle was vigorously attacked by other physicists, among them Einstein, who claimed that "God doesn't play dice with the world." For many years Einstein and Bohr carried on a famous debate, with *gedanken* experiments flying back and forth to refute or verify the uncertainty principle. A notable example of this is the paradox of Einstein, Rosen, and Podolsky, which required Bohr to redefine physics.

Einstein's "dice" objection is not valid from a religious point of view because God transcends the physical laws of the world which He created. Looking at it naively, one might welcome a probabilistic approach which implies that the miracles of the Torah are readily possible, albeit with a small probability. This implication is fallacious, however, because a miracle is a direct *intervention in* and *contravention of* the laws of nature. Assigning a small but finite probability to the occurrence of miracles takes them out of the realm of Divine intervention and into the realms of the physical laws of nature. The net result is the same as with the deterministic mechanics of the nineteenth century, i.e., it denies that the Creator is playing a role in His creation.

Let us then propose something quite radical. Perhaps the uncertainty principle, at least as it was viewed for many years, is wrong! This almost heretical viewpoint has actually been seriously considered by some prominent quantum field theorists during the last twenty-five years. Their considerations involve the area of hidden variables.

## HIDDEN VARIABLES

"Hidden variables" means that there are subquantum level particles which are very rapidly fluctuating. The $\psi$ functions are the statistical averages associated with these rapidly fluctuating particles, similar to the fact that temperature and pressure are the statistical averages of rapidly fluctuating Brownian motion of the particles in a gas. In this representation, as formulated by D. Bohm, the $\psi$'s are "real" physical entities and not just wave functions associated with the probability of a particle.

The indeterminance principle is still valid but becomes redefined in the following way. It relates to an indeterminacy in position and momentum resulting merely from a statistical fluctuation in the function, just as any statistically averaged function exhibits fluctuations. The particular relationship $(\Delta x)\ (\Delta p) \geq \hbar$ is not an inherent property of matter but a statement about a statistical average which is valid only in a particular regime, i.e., the regime in which quantum mechanics is valid. If one were to go down to a smaller level and look at the individual submicroscopic particles, most likely another, less restrictive, indeterminance relationship would be valid. The reason for invoking hidden variables is because the linear Schrödinger equation has problems when one goes to very small distances ($<10^{-13}$ cm) and very high energies ($>10^9$ eV). Singularities and infinities occurring as not necessarily mathematically rigorous solutions and renormalizations are required. As a result, particle physicists are looking at nonlinear Schrödinger equations, e.g.:

$$(-\hbar^2/2m)\nabla^2\psi + V(\psi^2)\psi = i\hbar\ \delta\psi/\delta\tau \qquad (3)$$

The solutions to these equations are very localized solutions which have the properties of particles. In this formulation the $\psi$'s are very localized solutions which look like particles, and they (the $\psi$'s) are in turn the statistical averages of the hidden variables. These very localized solutions are called "solitons." There is an analogue here to hydrodynamics, where shock waves (called solitary waves) can propagate over very long distances without dissipating. This is due to a balancing out of the dispersive nature of the wave with the nonlinear (velocity versus pressure) characteristic of the medium. In equation 3 the first term on the left is the dispersive term, and the second term provides the nonlinear balancing. In general, just as in hydrodynamics, two solutions are simultaneously possible—a very intense shock wave and a very low level sound wave. Similarly in equation 3 a solution of the form

$$\psi_T = (\psi_{soli.} + \psi_{wave}) \qquad (4)$$

is sought, where the first term on the right represents the "particle" and the second the wavelike nature associated with it. Using this formalism, the triumphs of the uncertainty principle (e.g., in explaining quantum mechanical tunneling and the diffraction pattern obtained with two slits) can be understood without invoking the uncertainty principle.

In the case of tunneling, where a particle is in a potential well, since there is a statistical fluctuation associated with the particle, there is a "probability" that it will jump over the barrier and "be" on the other side. Similarly, the pure wave component ($\psi$ wave) can tunnel through using *classical mechanics* as in *attenuated total reflection* merely because of its wave nature, without invoking quantum mechanics.

The result of all this is that an indeterminacy principle which implies a basic uncertainty in nature is no longer necessary. It merely reflects a statistical fluctuation.

As an aside, it is appropriate to comment here on Heisenberg's attempted philosophical extension of the uncertainty principle. To paraphrase: "The philosophical implications of physics are too important to be left to physicists." More generally, science is a methodology and not a philosophy. It tells the "how" and not the "why." In attempting to formulate a philosophy of science, one must be very careful lest one generate a religion (cult) out of science.

## RELATIVITY

Let us now consider the general theory of relativity (GTR). The basis of GTR is the theory of equivalence proposed by Einstein, viz., there is no difference between the forces in an inertial reference frame (e.g., under constant acceleration) and the forces in a local uniform gravitational field. The prime example is the elevator moving with constant acceleration and the elevator standing still in a constant gravitational field. The prediction of the bending of light rays by gravity follows.

The result of this formulation is that the very geometry of space is changed. The gravitational metric ($g_{\mu\nu}$) is introduced.

The shortest distance between two points for light to travel is no longer necessarily a straight line but a curved geodesic. In the interpretation of some physicists gravitation becomes (through the metric) a property of space.

The GTR has been successfully proven a number of times. Most dramatically, in the 1919 eclipse it correctly predicted the bending of the light of certain stars by the gravitational field of the sun. Here we would like to consider its applications to cosmology.

When Einstein originally postulated GTR, he believed in a static universe concept. However, measurements of the light coming from very distant galaxies in the 1920s changed this concept. The light exhibited a marked red shift. Red shifted light was observed from other astronomical objects which could be explained by relativistic effects. The measurements of the 1920s, however, indicated a correlation between distance and amount of red shift. The further distant galaxies exhibited a greater red shift. The red shift was thus explained as resulting from the Doppler Effect (shift toward longer wavelengths) due to these galaxies moving away from the observer. This naturally led to the hypothesis of an expanding universe.

## THE EXPANDING UNIVERSE

Using GTR, the expanding universe has a yardstick associated with it, $R(t)$, which is a measure of the expansion as a function of time. In addition, depending on the mass density of the universe, an infinitely expanding, unclosed universe is possible, or, if the mass is high enough, the universe is finite and closed in upon itself. The closure is an extension of the previously mentioned curvilinear metric which results from gravity. If there is enough mass, then there is enough gravity to close the universe in on itself. In this situation the 3-D universe is on the surface of a 4-D entity which is expanding. The universe is hence finite but unbounded—just as the surface of a sphere, a 2-D surface on a 3-D object, is finite but unbounded. In any event, in the finite universe case the universe would eventually reach a maximum expansion and would then begin to collapse. It is important to recognize in either case that since it is the metric that is expanding, one doesn't really notice locally, since everything is getting bigger together. This implies that not only is the mass expanding, but the actual fabric of the universe is expanding. This implies that there is something external to the universe in which the expansion is taking place: another realm of existence perhaps operating by different physical laws! Since the universe is a space–time continuum, this implies something outside space and time.

Current findings on the total mass of the universe indicate that there is not sufficient mass to form a finite, closed universe which will collapse upon itself. This contradicts other findings, and astrophysicists are busily looking for the missing mass. Their most likely candidate is neutral intergalactic hydrogen.

## BIG BANG

The expanding universe model led to the development of the Big Bang theory of cosmology. That is, at some distant time all of the mass of the universe was compacted together at some very high pressure and temperature. Then this mass exploded and as the density and temperature decreased the elements, galaxies and so forth were formed.

The expansion is continuing, however, as evidenced by the red shift. The matter and energy were initially in thermal equilibrium. It can be shown that the energy flux emitted by the radiation as it cooled should obey black-body statistics. It can further be shown that the product $R(t)T\gamma$ ($T\gamma$ = temperature of photon flux) is constant. Therefore, if one can measure the temperature of this radiation *now*, it is possible to go back by scaling to the "original" temperature, or vice versa. The temperature which this radiation should now have, based on estimates of $R(t)$, is around 3° K. Therefore, a verification of Big Bang would be the discovery of a radiation which is uniform, is isotropic, has a temperature of about 3° K, and has a black-body distribution. This discovery was made in 1965 (and beyond) by Drs. Penzias and Wilson, for which they subsequently received a Nobel prize.

## CREATION

From a religious point of view, most of the aforementioned is quite satisfactory. We have established

1. A creation (beginning) which implies a Creator.
2. A non-universe plane of existence beyond our own which similarly implies a Creator.

The nontrivial problem (from a religious point of view) which remains is the time estimates ($\approx 10^{10}$ years) which result from these studies for the age of the universe. The response to this is as follows: the Creator transcends space and time, and certainly we can accept that since the physical laws are His creations, he is certainly not bound by them.

Space–time is a continuum. For the Creator, Who transcends this continuum, the whole continuum (past, present, and future) are one. Therefore He can start it (create it) at any point that He chooses. This does not necessarily occur at what we perceive as a beginning. This means to say that the Creator could create the universe in such a fashion that an instant after this occurred if a person were on the scene (and was able to make sophisticated measurements), he would look into the earth and see geological strata that he would say took eons to form. He would look into the sky and see galaxies receding. In other words, an instant after time (as we now know it) begins, a whole universe exists, which—according to the physical laws in existence after creation—would have taken ten billion years to be formed. But this does not necessarily mean that this actually occurred ten billion years ago.

Why the Creator should choose to do this is certainly a valid question. Perhaps this is because He has given us free will. If it were readily apparent that the universe did not evolve in accord with physical laws, then the existence of the Creator would be very obvious and it would not require an act of faith to accept Him. The Creator, so to say, is hiding Himself through nature.

## CAN WE FIND THE CREATOR THROUGH STUDYING HIS CREATION?

We have attempted here to show that modern physics is not in conflict with belief in the Creator. It in fact seems to support, possibly, a belief in the Creator. Let us see if we can find, by looking at nature, a more positive response.

The Tzemah Tzedek, one of the early hasidic masters, writes that it is possible to know God the Creator through rational contemplation of the world. It is not necessary, he said, to invoke belief, which by definition transcends the rational faculties. This is very similar to the teleological argument that one can perceive God by seeing the order of the universe. (There is a subtle difference, however. The teleological argument states that we can *deduce* God's existence by observing the order of the universe, whereas the Tzemah Tzedek states that we can actually *see* God in the workings of the universe.) According to the viewpoint of the Tzemah Tzedek, a scientist who has a clearer perception of the natural world should more easily be able to perceive God. I would like to give a particular example of this and then discuss briefly the teleological argument in general.

Let us consider a very common substance, found all around us, vital to our existence: water. Water possesses a property which only a few other substances possess: it is most dense not in the solid state but in the liquid state, just above freezing, around 4°C. Almost all other substances which we know are most dense in the solid state. As a result of this special characteristic of water, life as we know it can exist on Earth. If water were like other substances, then, as the temperature drops, the denser water would sink to the bottom of oceans, bays, and lakes and freeze at the bottom because there, in its frozen state, it would be most dense. Bodies of water would freeze from the bottom up. Year after year the level of ice on the bottom would gradually get higher and higher, barely melting at all. Fish and other sea life would perish. Certainly this would make life impossible over most, if not all, of the Earth's surface. But as water is most dense in its liquid state, it freezes only at the surface, allowing the water underneath to flow freely. This most unique property, which allows life as we know it to exist on Earth, would seem to be a clear indication of the hand of a most beneficent Creator.

## DISPROVING THE TELEOLOGICAL ARGUMENT

Let us consider now the teleological argument in general. This states that from observing the order and apparent purposefulness of the universe one can perceive the Creator. Kant and his disciples argued against this, saying that positive purpose cannot be ascertained since

1. We find in existence things which look as if they have no purpose and happen by chance.
2. We cannot observe the whole of existence and we cannot traverse beyond the limit of the given world.

The response to this is in the negative proof, which has been formulated recently but was alluded to also in philosophic writings centuries ago.

The negative proof says that it is unreasonable to believe that a *complex, organized* system happened by chance. As Leibnitz said: "No man in his sane status would believe that a rational book was printed by throwing letters into the air." The popular probabilistic "philosophical" interpretation of quantum mechanics (already discussed here), however, shows us that this is actually exactly what many people wanted to believe!

What is the probability of a universe coming about by chance? The number of particles in the universe is somewhere around $10^{30}$ to $10^{40}$. The different ways that $N$ particles can combine is $N!$, that is to say, $N(N-1)(N-2) \ldots (2)(1)$. Therefore, the probability that they combine in a particular way is $1/N!$ For $N = 10^{30}-10^{40}$ this is a fantastically small number, possibly small enough so that the probability would not come to pass even in an almost infinite time, certainly not in the ten billion years mentioned earlier from the cosmological models. Hence we can safely say that the negative proof is correct. A reasonable person could not believe that the complex ordered structure of the universe came about by chance.

With respect to Kant's objection, the following can be said: If one were to read a book of wisdom which indicates the brilliancy of the author's mind but which contains some unintelligible words on some of its pages, it would appear that the unintelligible words had been written at random. Would anyone say that just because of these few apparently unintelligible sections the whole book happened by chance?

Concerning Kant's second point, scientists have now, indeed, been able to observe a very large part of the universe, extending from the very small to the very large, and even though there is much we don't understand, it seems clear that order and harmony exist throughout the universe.

Parenthetically you might then expect scientists to be on the forefront in extolling belief in the Creator! Indeed, we can find some great scientists for whom this is the case—for example, Einstein, who discussed an "omnipotent, just and omnibeneficent, personal God" in his essays.

It is perhaps true that we cannot perceive the purpose of Creation merely by observing order in the universe, but since by the negative proof we must accept that the world was created by a rational being, the Creator, then it follows axiomatically that the Creator has a purpose. What this purpose is we perhaps cannot fathom on our own. But once we accept Creation, then we must accept a purpose. It is not the role of science to describe this purpose. (As we remarked earlier, science merely tells *how* but not *why*.) This is the province of religion, and these are the complementary roles that science and religion play. In the words of Einstein:

> Now, even though the realms of religion and science in themselves are clearly marked off from each other, nevertheless there exist between the two, strong reciprocal relationships and dependencies. Though religion may be that which determines the goal, it has, nevertheless, learned from science, in the broadest sense, what means will contribute to the attainment of the goals it has set up. But science can only be created by those who are thoroughly imbued with the aspiration towards truth and understanding. This source of feeling, however, springs from the sphere of religion. To this there also belongs faith in the possibility that the regulations valid for the world of existence are rational. That is comprehensible to reason. I cannot conceive of a genuine scientist without that profound faith. The situation may be expressed by an image: science without religion is lame, religion without science is blind. . . .

I do not necessarily agree with Einstein's specific theology, but I do feel that his point is well made.

It is often lamented that there is no connection between modern science and religion. This is unfortunately a statement based on the confusion of science with the scientist. Science itself clearly proclaims the connection of the Creator to the created. The more we learn, the more this becomes apparent. Scientists who remain strangely unaffected by all this perhaps do so out of unwillingness to accept the obligation and responsibility which acknowledgment of the Creator entails. I feel, however, that this responsibility cannot be denied. In the words of the present Lubavitcher Rebbe *shlita*: "We all have a responsibility to enhance the spiritual content of the world through all our activities. . . . Science, which studies the very nature of the handiwork of the Almighty, must therefore also be used to enhance the welfare of mankind and to increase its awareness and desire for spirituality."

All of us, whether we are in the sciences or the humanities, share this burden. Just as we can all agree that it is *good* and *right* to use science and technology for creative purposes but bad and wrong to use them for destructive purposes, so should we see that to try to use science to hypothetically take the Creator out of the creation is destruction—for this destroys the very fiber, spirit, and soul of man.

From *B'Or Ha'Torah* 2, 1982.

# 7

# The Role of the Observer in *Halakhah* and Quantum Physics

*Avi Rabinowitz*
*Herman Branover*

The idea to write an article exploring the parallels between quantum physics and *halakhah* (Jewish Law) was inspired by a discourse of the Lubavitcher Rebbe on *Shabbat Parashat Shemini* in 1985, in which the dependence of physical reality on testimony of witnesses was discussed. The second author of this paper was present for the above discourse, and the parallels between the ideas discussed by the Rebbe and the views of quantum physics struck him. He shared his impressions with the first author and that is how work on this article began.

According to the philosophy of quantum physics, actual physical reality can exist (in the scientific meaning of the term *existence*) only as a result of measurement. When not being measured, the universe is in a quasi-real state amenable to description only in terms of probabilities and not facts. Some eminent physicists argue that the measurement must be performed by a conscious entity. According to this view, it is only measurement performed by a conscious being which can bring the universe into full reality.

This paper explores various fundamental perspectives in Jewish philosophy and *halakhah* that reveal parallels to this quantum world-view. These parallels pertain to ontology, which addresses the essence of existence, as well as to epistemology, which deals with criteria for knowledge. In the first section, we present a very brief summary of the relevant aspects of quantum theory. In the second section, we discuss some relevant aspects of Jewish philosophy and *halakhah*. These include the Jewish view of the Creation of the universe; the relation between the Creation, man's purpose, and man's unique characteristics; and finally the halakhic methodology for the establishment of fact. All of these are related by the common theme of determination of reality via observation performed by a consciousness.

## SECTION ONE: QUANTUM PHYSICS

For many years there was a debate among scientists regarding the nature of light. Certain phenomena seemed to indicate that light is a wave, while others pointed to its having a particle-like nature. The debate was stilled early this century when it was realized that all entities can be considered to be both waves and particles—some physical conditions cause a manifestation of the wave-like properties of these entities, and some cause a manifestation of their particle-like properties. This duality became one of the fundamental concepts of the newly developed quantum physics.

The duality concept arises as follows. If one were to eject an electron from an electron gun and run it through a slit of some sort, one would expect that since an electron is a particle, it would continue straight in a flat trajectory and reach a screen or a detector directly opposite to where it emerged. On the other hand, if one has a water or light wave, the waves bend as they traverse an opening, and form a complex pattern on the other side. When water or light waves are passed through a double slit, it has been shown that the resulting pattern can be attributed to the fact that each wave passes through both slits. It was shown experimentally, however, that even particles, such as electrons, form the same patterns as do waves, as though each particle passed through *both* slits—which seems impossible to our intuition. This led to the understanding that all entities in the universe are both waves and particles.

## PROBABILISTIC DETERMINISM

Prior to the advent of quantum physics, science believed that every event in the universe occurred as an inevitable and necessary result of previous events. The state of the universe at any one instant was believed to be totally determined by the states of the universe in the past, and in turn the present state totally determined what all future states of the universe would be.

Quantum physics, introduced at the beginning of the twentieth century, brought with it a drastic change in this viewpoint. Each event was now understood to be able to occur in a number of ways, with the actual way that it does occur left to "chance." Despite the fact that chance "ruled" each individual event, however, when numerous similar events occurred, the pattern that emerged resembled the results one would have expected using pre-quantum physics.

That is, each individual event, though occurring "at random," nevertheless contributes somehow to a pattern, in the aggregate, which can be determined beforehand. This combination of random and determined behavior is called here *probabilistic determinism* or *probabilistically determined randomness.*

Physicists were troubled, however, for this "random" aspect seemed to be contrary to the "spirit of physics."

Much inquiry was directed to the question of whether the chance aspect was only apparent—i.e., due to our lack of sufficient scientific knowledge and adequate instrumentation—or if it was an actual physical requirement. With the accumulation of more experimental evidence it was shown that, indeed, the nature of the physical universe appears to be such that intrinsically, at the most fundamental level, events are probabilistic and not deterministic.

## REALITY AND MEASUREMENT

The mathematical description of a quantum system is that of a wave which corresponds to the system being in all possible states simultaneously—a "superposition of states." (The physical interpretation of this wave is that of a probability distribution for the result of an individual event in an ensemble of cases or particles.) For example, in the case of the slit, each particle is represented by a wave function that corresponds to its going through both slits at all possible angles. However, when we look about us we always see unique states—for example, the unique impact point of the electron on the screen behind the slits.

Even the most subtle measurements on the particle while it is in flight, to determine its exact path, have been shown experimentally and theoretically to so disturb the path that the pattern is lost. Wheeler's delayed choice experiment has shown that one can even perform the measurements after passage through the slit, and retroactively affect the outcome. Therefore one cannot attribute a unique "physically real" path to the particle, even in theory—it is in a superposition state.

Thus there is a vast difference between the system prior to measurement—it is in a superposition of all possible states—and after measurement, when it is in a unique state. It can only be the act of measurement itself that causes this drastic change. It is counterintuitive because our observations of the universe about us are measurements, and we therefore always see only unique states, not superpositions.

Prior to its measurement, an event can occur in a number of ways and actually does so in some sense. Without measuring the state of a particle, we cannot say, "It is in some particular state, which is, however, as yet unknown to us." It is *not* in some particular state. Instead, it is (in some sense) simultaneously in *all* the possible states in which it can be! Indeed, even after measurement is made, and the particle is found to be in a particular state, we cannot then say that particle was in that state all the time—no! It had no definite state until one measured it. Our measurement forces the universe to assume one definite state from among all the possibility-states it is in prior to the measurement! This is usually called "the collapse of the wave function into reality."

This surprising, even bewildering, property can be interpreted as saying that the universe can emerge into reality only as a result of its measurement! In a way this is trite and in a way it's radical. It is trite in the sense that for at least two centuries philosophers such as Berkeley and the positivists have explored the idea that reality is only set by our consciousness of it, because we can perceive things through the senses. It is radical, however, in that this result has now been achieved by physics. Therefore, what is true in the realm of words and ideas has been shown by physics to be true of physical reality as well: the actual thing itself is not set until it is measured.

## Summary

From the preceding discussion, we can make the following summary: All entities in the universe have a dual nature, one material and localized (particulate) and one nonmaterial and nonlocalized (wavelike). The universe, or any subsystem of it, is capable of being in two (or more) mutually contradictory states simultaneously. Saying that the universe is uniquely in one state is just as invalid as saying that it is in the other state. However, a decision can be made between the two by having a measurement made of the situation. Both views are partially correct, but only until the measurement is made. After that, only one becomes correct. However, this does not imply that it was always the correct one; there *was* no "correct" state until the measurement was made.

## THE ROLE OF CONSCIOUSNESS

What is the active factor in a measurement which causes this emergence into reality? According to some leading physicists, this factor is consciousness. The great mathematician John Von Neumann, who provided a rigorous mathematical foundation for quantum mechanics, believed that only a *human* consciousness can collapse the wave function.[1,2]

The eminent Nobel prize–winning physicist Eugene Wigner writes:

It follows that the quantum description of objects is influenced by impressions entering my consciousness. . . . It follows that the being with a consciousness must have a different role in quantum mechanics than the inanimate measuring device.[3]

The famous physicist John Wheeler has taken this one step further. According to him the entire universe can emerge into true physical existence only via the observation of a consciousness!

[Perhaps] no universe at all could come into being unless it were guaranteed to produce life, consciousness and observership somewhere and for some little length of time in its history-to-be? . . .

[T]he observer is as essential to the creation of the universe as the universe is to the creation of the observer. . . .

[T]he universe would be nothing without observership, as surely as a motor would be dead without electricity. . . .

[I]s observership the "electricity" that powers genesis? . . .

"[O]bservership" allows and enforces a transcendence of the usual order in time. . . .[4]

Thus, according to "quantum metaphysics,"[5] a consciousness is indispensable to the universe if it is to emerge into reality. Physical reality can be said to exist only as a result of our presence within it or, more precisely, as a result of our perception of it.[6]

Wheeler has constructed a fascinating diagram to illustrate this concept (see Figure 7–1). Explaining this diagram, he writes: "Beginning with the big bang, the universe expands and cools. After eons of dynamic development it gives rise to observership. Acts of observer-participancy in turn give tangible reality to the universe not only now but back to the beginning."[7]

## FREE WILL, QUANTUM PHYSICS, AND THE COLLAPSE OF THE WAVE FUNCTION

A free-willed decision, in order to be truly free, has to be unconstrained by the laws of nature and not determined by any physical phenomena. Hence free will must be neither the result of deterministic processes nor the result of random processes occurring in accordance with the natural order of phenomena. Hence if the universe contains a free will, this free will must operate via interactions which transcend both the determinism of classical physics and the randomness of quantum physics. Free will is then unique in this respect.

If some entity exists which can collapse the quantum wave function, then it is reasonable to postulate that this entity has to be free will, since, as we just discussed, only free will transcends quantum randomness, as it transcends nature in general.

Since a consciousness can affect the universe only if it has a free will, and a free will is by definition unthinkable without a consciousness, we will assume in the course of further discussion that free will subsumes within itself the concept of consciousness.[8]

## EXISTENCE OF THE UNIVERSE AND THE ROLE OF FREE-WILLED CONSCIOUS BEINGS

We will try now to apply the conclusions we reached above to the question of the existence of the universe. As we saw, according to quantum metaphysics, the universe can emerge into reality only when it is observed by a consciousness. This consciousness possibly must function in a nonquantum fashion in order to "collapse the wave function." The only such nonquantum factor in the universe is free will. Thus, we postulate that it is the presence of a free-willed, conscious being which enables the universe to emerge into reality!

## SECTION TWO: HALAKHAH AND PHYSICAL REALITY

According to Jewish thought, the true reality is the spiritual realm, the physical cosmos being God's precision-crafted instrument for achieving spiritual goals. Indeed, the physical universe is a shadow of the spiritual world, the illusion perceived by limited beings who are in contact with the spiritual cosmos but can directly sense and perceive only its shadow. Human free-willed moral choice connects the two realms, and this moral activity gives meaning to the existence of the universe.

Thus, since the true reality of the universe is the spiritually meaningful aspect, it should not be surprising that the emergence of the universe into reality is so intimately bound up with the emergence of those beings who endow physicality with meaning. Furthermore, once this connection is understood, it is most appropriate that the very characteristics of man which allow the emergence of the universe into reality (i.e., his free-willed consciousness) are the very same characteristics which endow it with meaning. We can thus see the fundamental interrelationship between meaning, purpose, free will, consciousness, and the very nature of reality (and how this is reflected in Creation).

### Summary

We have seen that according to quantum (meta)physics, reality is established via the observation of a (free-willed) consciousness. In addition, we have shown that according to Jewish thought, free-willed choice gives the universe meaning and is thus the "motivation" for the very existence of the universe. We will now explore how the nature of everyday Jewish law is correlated with this radical conception of the reality of the universe.

## MAN, TORAH, AND REALITY

According to Jewish tradition, the Torah is the "blueprint" which was used by God in creating the universe.[9] Moreover, God is continually renewing Creation—re-creating the universe every moment—by interpreting the blueprint for the re-creation of the universe at that instant. These statements pertain to a more abstract form of Torah, to a Torah that exists spiritually in the form of "black fire on white fire."[10] However, once this Torah was translated into human terms and given over to man at Sinai, the prerogative of its interpretation and application rested solely with man; it is man's responsibility and his alone. And through involvement with Torah, man actually becomes a partner in the continuous renewal of the cosmos.

In our universe, the operative concepts are "free-willed conscious choice," not "nature," i.e., objective scientific fact. It is the former which provides the motivation for the existence of the latter (the purpose of the universe is free-willed choice), and it is the former which causes the latter to emerge into existence (collapse of a probability wave). Thus, it will not come as a surprise that, in Jewish law, free-willed conscious moral choice rules over the "laws of nature," rather than vice versa.

The true reality is the spiritual one. The physical is in existence only to serve the spiritual. The entire physical universe is an artifact created by God. Man is a precisely crafted instrument designed to interact with the physical universe in ways which have the potential to achieve spiritual goals unattainable without the vehicle of the physical. Thus the physical is of central importance, but only as a means: It can have sublime beauty and dignity, but only by virtue of its ability to achieve in the spiritual realm. Thus the human body, rather than being a hindrance to spirituality, is a potentially holy physical tool which can control the spiritual.[11]

Indeed, every action/thought/word affects the spiritual cosmos, and the Jewish way of life as prescribed by the Torah is designed to resonate with the spiritual and to correctly utilize the physical in order to elicit the fusion of ultimate spirituality with the physical. God is the Creator of the universe and of man. He has designed the universe, man, and the way to complete each other, to complement one another, in a self-consistent optimum system. He communicated the method to man in the form of the Torah (Written and Oral), and He, the Designer and Creator of this system, has promised us that we are superbly qualified to successfully perform our role and attain our goal. More than this, we are told that since we are designed to be the best instrument to achieve this goal within the context of a physical universe, when we perform at optimum, we can determine the nature of the physical/spiritual interaction, and therefore of *halakhah*.

## HALAKHAH AND THE
## DETERMINATION OF REALITY

There are three ways by which man determines halakhic reality: by determining which one of several rabbinic opinions shall be authorized for implementation; by determining which of several scenarios (having halakhic relevance) is the factually correct one; and by determining the reality itself, i.e., forcing the reality to conform with the halakhic opinion.

## HALAKHAH AND REALITY

1. *Halakhic decisions:* Normally, one would assume that of two conflicting viewpoints regarding legal matters in a Divinely mandated code of life, one must be incorrect. However, the power of man to determine spiritual reality is reflected very clearly in halakhic philosophy, and in *halakhah* itself. According to the Talmud, when two qualified rabbis arrive at conflicting and mutually exclusive conclusions, if both have researched the matter as required, and both are sincere and interested only in the truth, then both are right. This attitude is recorded in the Talmud and is given here in rough translation: "For two-and-a-half years the schools of Rabbi Hillel and Rabbi Shammai were in dispute regarding a certain ruling. Finally, a voice came from Heaven and declared: 'Verily, both views are the words of the living God, but the ruling is according to Rabbi Hillel.'"[12]

Since the spiritual reality is determined by *man*, there is no paradox involved in two mutually contradictory results being both right, as long as they represent two human opinions arrived at through religiously legitimate means. The Torah also gives us rules for deciding the actual path to take when we are presented with two or more valid opinions, for example, the principle of decision by majority.[13] Once one of the paths has been accepted by rabbinical authority, it alone becomes the *only* viable path of action. Even though the other path is still theoretically correct, it becomes absolutely unacceptable as an actual path as soon as the religious reality is decided in favor of the other.

For example,[14] the dates for Jewish holidays were set according to rabbinical ruling based on observation of the moon and on calculations. One year the date for Yom Kippur (the holiest day of the year) calculated by Rabbi Yehoshua differed by one day from the date arrived at by Rabban Gamliel. Since Rabban Gamliel was the *nasi* (chief justice), the ruling followed his view. In order to prevent anarchy, he then ordered Rabbi Yehoshua to come to see him on the day that Rabbi Yehoshua considered to be Yom Kippur—and to do so in a manner forbidden on Yom Kippur. This would prove that there was only one authoritative ruling. Rabbi Yehoshua came, violating what he considered to be the holiest day of the year, in deference to Rabban Gamliel's view. Rabban Gamliel then told him that he considered Rabbi Yehoshua to be his superior in wisdom and learning, but that as *nasi* his word was nevertheless law. Thus, even though Rabbi Yehoshua's view was possibly more logically valid than that of Rabban Gamliel, as soon as the decision was made in favor of the other view, his own ruling became totally unacceptable as religious law, and the day he considered the holiest ceased to be so.

We have just seen that two conflicting opinions can both be considered valid, yet that only the majority opinion is accepted as normative. This is an example of the simplest level at which man determines the halakhic reality, i.e., that of determining which of several opinions shall be accepted as normative *halakhah*. We will now present examples of the other two levels on which this is true: disclosure of factual reality and causation of factual reality.

2. *Disclosure of factual reality*: In the same *mishnah* discussed above, Rabbi Akiva brought proof from the Torah that even if the court set a festival on the incorrect day through its own error, the declaration was religiously valid. The following *mishnah* relates that even if all the people of Israel, including all the court justices, saw the new moon and/or received the testimony of the witnesses, if the court did not *declare* it as the new moon (by the day's end), the new moon was considered to begin the next day.

Even when *halakhah* seemingly depends on objective physical reality, such as the temporal sanctification at the time of the new moon, what actually determines the *halakhah* is the decision/measurement made by man (in this case, the court ruling), even though the physical facts seem to present a contradiction. Neither nature nor supernatural events can usurp the prerogative of man in deciding *halakhah* and thus the spiritual reality. This can be seen graphically in the account presented in the following narrative from the Talmud.[15]

Rabbi Yehoshua and Rabbi Eliezer were arguing a point of religious law. Rabbi Eliezer was unable to convince Rabbi Yehoshua of his point, and so he then tried to convince him by using miracles. He said: "If I am right, let the stream run uphill." And it did so. This, however, did not convince Rabbi Yehoshua. "If I am right, let the wall bend," declared Rabbi Eliezer, and the wall indeed bent. Rabbi Yehoshua, however, remained unconvinced! So Rabbi Eliezer called upon Heaven itself to decide the matter. He said, "If I am right, let Heaven prove it." Indeed, a voice came from Heaven and declared, "What do you want from Rabbi Eliezer? The ruling is always in accordance with his views." At this point, Rabbi Yehoshua arose and quoted the Torah saying, "It (the Torah) is not in Heaven."[16] By this he meant that it is man—not physical law or even Heaven—who decides spiritual reality. That is, he did not dispute the fact that Heaven had decided in favor of Rabbi Eliezer (and that probably this meant that "nature" favored him as well); rather, he claimed that Heaven had already delegated the authority in such matters to man alone!

The most far-reaching example of such a philosophy can again be found in the Talmud.[17] There we are shown quite clearly that neither the inhabitants of the spiritual realm nor God Himself defines *halakhah*; only man can do so. A debate about a religious ruling was taking place between the members of the "Heavenly College" (the souls of deceased saintly rabbinical scholars) and God Himself: God held to one view and the members of the College to another. The College then decided that in order to reach an authoritative ruling, neither God nor Moses could decide, but rather that the living rabbinical authority (Rabbi Bar Nachmani) must decide.

The Talmud also tells us what God's view of all this is: At the end of the case with Rabbi Yehoshua and Rabbi Eliezer described previously, Elijah the Prophet is quoted as reporting that God was immensely pleased with man's understanding of his role in determining religious reality, and that He even laughed with joy, saying: "My sons have bested me, my sons have bested me." Thus on questions of *halakhah* and spiritual reality, the world, the Heavens, and God Himself all defer to man, though he be incomplete and limited.

3. *Halakhah and the Causation of Reality:* Since the Torah is the blueprint of Creation, man's interpretation of it can have cosmic effect; halakhic ruling causes physical reality to emerge and even change so as to conform with the blueprint from which it was created. Thus an extremely potent level on which man can determine physical reality is that of halakhic decision. This astonishing concept is referred to in a famous passage in the Jerusalem Talmud.[18] In order to understand the passage, one must be familiar with the concept of the leap year in the Jewish calendar.

A Jewish leap year involves the addition of a thirteenth month to the twelve-lunar-month year, in order to correlate it with the solar year. If a child is born in the middle of the last month of a non-leap year, then the month of birth is both "the twelfth" and "the last" month of the year. During leap years, when the twelfth month is not the last month, it is not obvious which month is the child's birth month: the last (thirteenth) month, or the twelfth (second to last) month. The talmudic ruling is that the birthday is on the last (thirteenth) month.

The passage in the Jerusalem Talmud which we mentioned above concerns physical changes in the body of a child involving a lessened ability of the body to regenerate certain tissues; this change is determined as occurring at the conclusion of the third year of life. Thus, if the child's birthday is, for example, on the fifth day of the last month, then in the third year, on the fifth of the last month, the physical change occurs. However, if at the end of the month, *after the change has taken place*, it is decided by the court that a thirteenth month will be added to the year, the third birthday according to *halakhah* has not yet occurred (it will occur only on the fifth day of this newly added thirteenth month), and thus the physical change should not yet have occurred. According to this talmudic source, the physical change will, in this circumstance, actually reverse itself in conformance with the halakhic decision! (If an observation is made already, the reality is "frozen" and cannot be changed retroactively by a later observation; it can, however, be changed if no observation is made.)

For another instance, the famous Rogatchover *Gaon* once indicated a similar philosophy regarding the ability of *halakhah* to determine physical reality. A woman once came to him with a defective slaughtered chicken to ask whether or not it was *terefah* (an animal with a terminal disease or fatal defect that renders it not kosher). The chicken seemed to be a clear-cut case of *terefah*, yet the Rogatchover's students noticed that their rabbi spent hours checking it, consulting books, thinking and so on, rather than spending the usual few minutes. They were astounded that this great genius would waste his time on such a trivial matter instead of simply buying the chicken from the woman and throwing it out. In the end, the Rogatchover was able to find a way to reach the halakhic decision that the chicken was not a *terefah*, i.e., disease was not fatal according to the halakhic definition of "fatal." The students were stupefied at the exertion of such genius over such triviality.

The next day, a woman came to the Rogatchover wailing and sobbing. Her husband was deathly ill, she cried, asking the rabbi for help. The Rogatchover asked the woman what was wrong with her husband. She described precisely the disease which afflicted her husband. The Rogatchover then answered thus: "Don't worry—only yesterday I examined a chicken with the identical disease and I ruled that it was not a *terefah*—your husband is therefore not deathly ill and will recover!"

The biblical sources given for this far-reaching concept are passages in Psalms and Isaiah, whose hidden meanings had been passed down along with the text from generation to generation. For example, in Isaiah 44:24–26, we read: "I am the Lord that makes all things; that stretches the heavens, alone; that spreads abroad the earth by Myself; that frustrates the omens of impostors, and makes diviners mad; that turns wise men backward, and makes their knowledge foolish; that confirms the word of His servant, and performs the counsel of His messengers."

God created the universe *ex nihilo* but has left to man the fashioning of this raw material, as indicated in the traditional interpretation of the passage: "For then He rested from His work, which He created to be fashioned" (Genesis 2:3).

The Sages interpreted the apparently superfluous term "to be fashioned" as indicating: "All that was created in the six days of Creation requires 'fashioning' (by man)" (*Bereshit Rabbah* 11:7).

## QUANTUM HALAKHAH

We can now compare the halakhic view of reality with that of quantum physics. According to the Talmud, two or many logically valid halakhic rulings may be consistent with the available facts. Indeed, two independent valid viewpoints can be diametrically opposed and mutually contradictory. *Halakhah* allows either one to determine the reality in principle but will accept only one of the possibilities in actuality.

The analogy between this conception and the idea of the multiple possibilities for reality prior to a "measurement" in quantum physics is now obvious. Although both are the (living) words/realities of (the living) God, only one can determine reality. Also, even though all entities can be measured to produce the characteristics of either waves or particles—both are living reality—only one aspect by itself can be seen "statistically" as the result of an actual measurement.

Furthermore, just as according to quantum physics (or metaphysics) nature has delegated to man the ability to determine the nature of physical reality within the limitations of natural law, similarly, according to Torah, God, the Creator of nature, has delegated to man alone the ability to determine the nature of spiritual reality, within the limitations of *halakhah*. Spiritual reality then influences the physical. Of the two levels, the physical is merely the means to the spiritual end. Thus the determination by man of spiritual reality is even more fundamental than his determination of physical reality. So, too, it is man's

spiritual qualities (free-willed consciousness) which are more fundamental than these physical qualities.

As illustrated in the previously cited instances, it is up to man to
use his own limited sense of right and wrong, guided by Torah criteria,
to determine the *halakhah* and thus to determine reality. *Halakhah* is
the guideline for the Torah way of life, the way to achieve one's purpose—and it is this purpose which also gives meaning to the universe.
Since it is man's consciousness and free will which invest his choices
with the possibility of meaning, it is therefore only free-willed consciousness which has the possibility of conducting reality-determining
observation and measurement.

Nature by itself is powerless to achieve self-realization; man is
required to bring both himself and the universe into reality. Thus
nature cannot determine *halakhah*. Similarly, God Himself does not
decide *halakhah*; it is man's prerogative and sole responsibility. Man,
alive and physical and yet spiritual as well, albeit limited and fallible—
or perhaps *because* he is limited and fallible—is uniquely qualified, by
virtue of his possessing a free-willed consciousness, to determine the
nature of physical and spiritual reality.

## SOME EPISTEMOLOGICAL CONSEQUENCES
## OF QUANTUM METAPHYSICS

As we have seen, quantum physics connects ontology (being)
with epistemology (knowing), and quantum metaphysics postulates
that the universe can emerge into true physical existence only when
there are (free-willed) conscious beings in it. According to this scenario, man is not a random product of the universe but is rather a
necessary condition for the very existence of the universe. In addition,
since the universe can emerge into existence only when free-willed
conscious man is present within it, there is no true physical reality to
any time prior to the emergence of the first free-willed conscious man.
According to the Torah, this man was Adam.[19]

Of course, other theories have been advanced regarding the
emergence of man. However, these by definition relate to a time prior

to the emergence of conscious individuals. According to the approach of quantum physics explored here, these theories cannot relate actual physical historical events since there were no such actual events prior to the emergence of free-willed conscious beings.

Whether or not one accepts the Torah as being able to provide information on matters inaccessible to scientific inquiry is in itself a matter of free-willed choice. However, if events "prior" to the emergence of the first free-willed conscious being are undefined scientifically, then it is only to be expected that there will be differences between the description made of the emergence of this first free-willed conscious being by a source limited by quantum physics—such as man and his theories—and the description made by a source originating outside of physicality—such as the Torah.

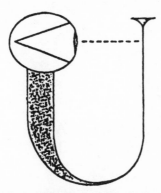

**Figure 7–1.** Physicist J. A. Wheeler's diagram illustrating the concept of "quantum metaphysics." Explaining this diagram, he writes, "Beginning with the big bang, the universe expands and cools. After eons of dynamic development it gives rise to observership. Acts of observer-participancy in turn give tangible reality to the universe not only now but back to the beginning." (See page 68 and note 7 in this chapter.) Used by permission of the author.

From *Fusion* (the Proceedings of the First *B'Or Ha'Torah* Conference, Miami, FL, 1987), published by Feldheim, New York-Jerusalem, 1990.

# 8

# Geocentrism

## Avi Rabinowitz

The Copernican revolution discredited geocentrism and thereby caused the onset of the Christianity–science "conflict." However, although certain biblical passages and a number of statements in the Talmud can be interpreted as implying a geocentric view of the universe, a Torah–science conflict around the issue of geocentrism never existed. Further, although Copernicus's system was widely viewed as having disproved the geocentric one, by using Einstein's general relativity theory one can show that the geocentric picture of the universe is no less correct than the heliocentric one. As the eminent physicist Arthur Eddington remarked, "[according to] scientific theory there is no absolute distinction between the heavens revolving around the Earth and the Earth revolving under the heavens; both parties are (relatively) right."

---

The author has expanded and updated this chapter for publication here.

# PART ONE: INTRODUCTION

To many people, science and religion seem to be incompatible. This problem is indeed a very old one. For about two thousand years, in each age, the science then believed to be valid was used by heretics in attempts to disprove the validity of the Torah. However, virtually all those arguments are today irrelevant since they are based on concepts outmoded by advances in science itself. The main problems relevant today have their origin in the more modern scientific discoveries of the last few hundred years.

In particular, the modern crusade to discredit the Bible had its roots in the Church–astronomy controversy over geocentrism. The medieval Church had adapted much of Aristotle's philosophy to its own theology and accepted the geocentric view of cosmology: the Earth was believed to be stationary and at the center of the universe, while the rest of the universe (the sun, planets, and stars) rotated about it.[1]

# HISTORICAL OUTLINE OF COSMOLOGY[2]

*Anaximander* (approximately 600 B.C.E.). Earth is a sphere suspended in space. The sky is spherical and concentric with the Earth.

*Pythagoras* (sixth century B.C.E.). Each heavenly body rides on a "sphere" in space. They are all concentric with the Earth but are at successive distances from it.

*Aristotle* (384–322 B.C.E.). Improved and systematized the scheme of Pythagoras. Provided a philosophy to explain why the spheres move. (They were considered to be living and bearing intelligence.) Gathered all the knowledge available at his time into a comprehensive unified system.

*Aristarchus* (third century B.C.E.). Earth revolves about the sun once a year and about its axis once a day. (This is Copernicus's system, eighteen hundred years before Copernicus! Aristarchus is now sometimes called the "Copernicus of Antiquity.") His system was not accepted, however, because (1) people could not conceive of the Earth moving, especially since no motion was perceived, and (2) Aristarchus did not work out the details of his theory and did not provide planetary tables based on it, so that no one could check the planetary motions to verify the theory. Aristotle's theory remained the accepted one.

*Ptolemy* (second century C.E.). Basing his work on studies done by Hipparchus (second century B.C.E.), created astronomical system by perfecting Aristotle's construction. Enabled accurate predictions of the motions of heavenly bodies. Wrote the *Megisti Syntaxis* (the *Almagest*). Gave many reasons to reject Aristarchus's system.

*Copernicus* (1473–1543). Possibly influenced by ideas of Nicetus of Syracuse and Aristarchus. Countered all of Ptolemy's arguments against a moving Earth. His astronomical system placed the sun at the center of the universe, fixed the stars on the motionless outer "sphere," and had the Earth spinning on its axis and rotating about the sun in a circle.

*Bruno* (1548–1600). Universe is infinite in spatial extent. The sun is merely one of its stars. Each star has its own planets, probably with intelligent beings on them. Religious "heretic." Burned at the stake by the Inquisition in Rome.

*Galileo* (1564–1642). Used telescope to discredit Aristotelian system and support Copernican theory. Explained how Earth moves without this motion being felt by its inhabitants (inertia). Imprisoned by the Inquisition at Rome.

*Kepler* (1571–1630). Three laws of planetary motion. Orbits are ellipses, not circles. Planets move at varying speeds during orbits. Made the first real sun-centered astronomical calculations.

*Newton* (1642–1727). Explained the structure of the solar system and its motions based on his law of gravitation.

*Einstein* (1879–1955). Relativity theory. Gravitation understood as manifestation of the effect of matter–energy on space–time.

Copernicus (1473–1543) showed that it was far easier to compute the orbits of the celestial bodies if it was assumed that the Earth orbited the sun (heliocentrism) rather than that the sun orbited the Earth (geocentrism). This was a mathematical result only and there was no actual proof that reality followed the mathematically simpler model. Of course Copernicus himself believed it was a true model of reality. However, an unsigned introduction to his major work (probably inserted without Copernicus's knowledge) presented the thesis as merely a mathematical convenience.

However, Galileo (1564–1642) used his telescope to present what he felt was *proof* that the Earth actually did orbit the sun. Thus Copernicus's heliocentric system was now presented as an actual model of the solar system rather than merely as a mathematical trick to simplify calculations.

The Church, insisting that the Bible states that the sun moves about the Earth and that the Earth is immobile, found itself in conflict with science. This conflict led to a long and bitter struggle between religion and science. Indeed, many Jews, following their Christian contemporaries, believe that the overthrow of geocentrism was one of the cardinal blows to religion in general and to the Torah in particular.

There are three ironies inherent in this belief:

1. It is truly ironic that all the scientific disproofs of geocentrism have in themselves been disproved by science. The development of Einstein's general theory of relativity has turned the tables and shown that the Copernican system, which supposedly disproved the geocentric one, is no more valid than the geocentric system itself. (This will be shown in detail in Part Two.)

Thus dogmatic statements claiming that astronomy had disproved the Bible have themselves been shown to be scientifically incorrect, and the original, prototypical "science versus Torah" argument has been invalidated.

2. The second irony is the belief that there ever was a real science–Bible controversy to begin with over this issue: It is manifestly clear that the very astronomers who found themselves involved in the struggle against the Church on this issue were themselves devout believers in the Bible and in its Divine origin. This is shown in the following citations which typify the respective personalities of the three great astronomers, Copernicus, Galileo, and Kepler:

About Copernicus Luther commented: "The fool would overturn all of astronomy. But in the Holy Scriptures we read that Joshua ordered the Sun to stand still, not the Earth."

Copernicus replied: "To attack me by twisting a passage from Scripture is the resort of one who claims judgment upon things he does not understand. Mathematics is written only for mathematicians. The statement that Joshua made the Sun stand still, and not the Earth, must not be taken as a revelation concerning Nature. The Bible contains no astronomy, not even the names of the planets. Of course Holy Writ cannot err, but some interpreters of it can. For example, it would be blasphemy to take literally the passages concerning God's wrath, hatred, and repentance; this everyone admits. Similarly, those Scriptural passages that do not agree with the findings of science are not to be taken literally. For the laws of nature operate with absolute inevitability, and these laws are the creation of God."

Galileo once exclaimed: "I know now what the silver girdle around the celestial sphere is; I am filled with amazement and offer unending thanks to God that it has pleased Him to reveal through me such great wonders, unknown to all the centuries before our time." Though Galileo's findings were condemned by the Church for contradicting Scripture, he himself believed in the Bible and simply understood the relevant passages as being meant allegorically, since they had to be written in the terms common to those who lived at the time the Bible was given to man.

Kepler was motivated to look for simple laws of planetary motion because he believed that the cosmos, as the work of a masterful Creator, was based on elegant yet simple patterns. He was deeply convinced that the spirit of God revealed itself most purely in geometry. After discovering his famous laws of planetary motion, he was overwhelmed. "Dear Lord," he prayed, "Who has guided us to the light of Thy glory by the light of nature, thanks be to Thee. Behold, I have completed the work to which Thou hast called me. And I rejoice in Thy Creation Whose wonders Thou hast given me to reveal unto men. Amen."[3]

Thus the issue was clearly a clash between scientists and theologians rather than between science and religion. Some scientists, including Galileo, even "derived" Copernicus's system from the Bible itself.[4]

3. Thirdly, it is ironic that many Jews feel that somehow the validity of the Torah was disproven—in actuality neither Jews nor Judaism were in any way involved. Indeed, the *Zohar* on Leviticus 10:1 discusses the rotation of the Earth.

> In the Book of Rab Hamnuna the Elder it is explained further that all the inhabited world turns in a circle like a ball . . . there is a part of the world where it is light when in another part it is dark, so that some have night while others have day. Also, there is a place where it is always day and where there is no night save for a very short time.

This indicates that the idea of the rotation of the Earth is not contrary to Jewish belief. In fact, this source was quoted by Rabbi Avital Sar Shalom of Barzila in his *Emunat Hakhamim* (about 1630) in support of the theory that the Earth is circular and rotates on its axis.[5]

David Gans, a student of both the Maharal and Kepler, wrote a book, *Magen David*, published in Prague in 1612, in which he praises Copernicus and his system. (This is the earliest known reference to Copernicus in Jewish writings.)

Joseph Solomon Delmedigo wrote in favor of Copernicus's work. Probably in reference to the Christian attitude to Copernican theory, he stated that attempts to deduce astronomical systems from passages in the Torah were misguided.[6]

Rabbi Jacob Emden (1697–1776) also supported Copernicus's ideas. In his commentary on the prayer book, he explained that the Hebrew word for Earth, *eretz*, is derived from the root *ratz*, meaning "runs," because of the rotation and motion of the Earth.

From the above, it is clear that the Copernican revolution was not viewed in any way as a threat to Jewish belief. On the contrary, the Copernican system was accepted wholeheartedly.

Science is far from anathema to Judaism. On the contrary, according to the Rambam (Maimonides), it can serve as the catalyst to a profounder observance of a fundamental Torah precept and to a deeper

understanding/experience of the Jewish Way. One of the most fundamental religious requirements is to love God and to fear Him. (Here,
"fear" is meant in the sense of "awe," rather than "being afraid of.")

The Rambam (in the second chapter of the *Sefer Ha-mada* in his
*Mishneh Torah*) states that the Commandments to love and fear God
should be fulfilled via the study of God's actions/creations. These
actions/creations are of course revealed to us as "nature." Thus, the
study of nature is a means of achieving love and fear of God. The
Rambam explains that a deep examination of nature reveals the infinitely deep and precious Wisdom of God which then leads to an over-
whelming feeling of praise and love for God. One who deeply studies
nature is not only *intellectually* aware of the infinite Wisdom of God;
he *feels* it also. Not only is he intellectually aware that an infinite
gap separates his understanding from the perfect understanding of
God but he *experiences* its consequences. He begins not only to love God
but also to be in complete awe of God. Indeed, study of nature can
lead to an appreciation of both the Infinite Wisdom of God and the
infinite difference between man and God; this can, in turn, lead to
the love and fear of God spoken of in the Torah.

Clearly, the mere fact that the Christian Bible incorporated the
written Torah within itself is no guarantee at all that issues of relevance
to Christian theology and to Church behavior are also of relevance
to Judaism. Nevertheless, the feelings prevailing in Christian society, to
the effect that the Bible had been disproven, somehow managed to in-
fect many Jews as well. Regardless of the fact that neither geocentrism
nor Torah were in any way disproven in actuality, the widespread
misconception that they were *disproven* has deeply affected the philosophical beliefs of many people.[7] Perhaps the major such belief to be
vitally affected is that concerning the significance of man. This issue
is dealt with in Part Three.

## PART TWO: GEOCENTRISM IN THE LIGHT
## OF GENERAL RELATIVITY

In Part Two we will present various ideas from the general
theory of relativity to demonstrate that, indeed, the Copernican

system is no more valid than the geocentric one. Readers who prefer to skip the more technical aspects can proceed directly to the
last few pages of this section or to the beginning of Part Three,
without losing the thread of the discussion.

Although the Earth does not occupy a physically unique position
in the universe, it would be entirely consistent with the laws of physics
to say that the sun revolves about the Earth, and that the Earth is the
center of the entire universe. To understand this, we must know something about general relativity, cosmology, and the philosophy of physics.

Before we go further, however, we must clarify what "center of
the universe" means! To do this, we must explain what we mean by
saying that something is "at the center." There are actually different
types of "centers," as follows.

*The Center of Rotation*

The wooden stick in Figure 8–1 is connected to the board be-
neath it by a nail which serves as a pivot. The stick rotates, as shown,
around the pivot. Thus the nail can be termed the "rotational center"
of the stick (or of the stick-board–nail system), even though it is certainly not the physical center.

Analogously, the universe may rotate about some object which
is not located at its center. Thus, the Earth might be the rotational
center of the universe without being its spatial center.

*The Center of Expansion*

Imagine that you are surrounded by four people. Each of the four
then walks away from you, each in a different direction (i.e., north,
south, east, west, as in Figure 8–2).

Thus when you look in each direction, you will see people moving
away from you. In contrast to this, if any of the other people look they
will see a quite different pattern. They see people only when they look
behind them, and thus the pattern they see cannot be symmetrical. This
difference is, of course, due to the fact that you, and only you, are the
center of this dispersion pattern. If we call this an "expanding" pattern,
then we can say that you are the "center of expansion."

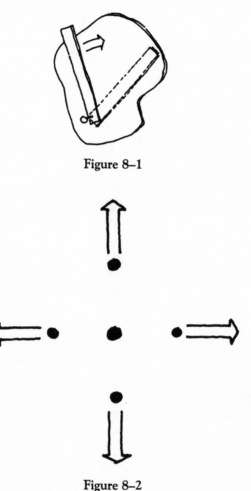

Figure 8–1

Figure 8–2

Now we increase the number of surrounding people to twelve (see Figure 8–3). We place them on a circle about you so that each is the same distance from you and then ensure that each one is separated from the other by a given distance. They then begin to walk away from you (radially), each with the same speed. Thus, wherever you look, you will see people moving away from you, with all of them doing so at the same speed; the pattern is very symmetrical, with you at the center.

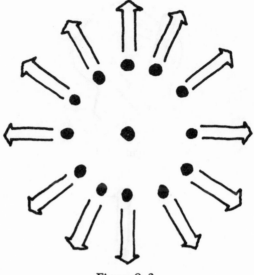

**Figure 8–3**

Of course, any of the others would see a totally different picture. They would see people only in the directions "behind" them and they would see everyone moving at different speeds. In fact, each one could probably look back at any time and figure out that it is *you* who are standing still, that *you* are the center of the pattern, and that everyone else (including themselves) is moving away from you with a constant speed.

Thus, whenever something is the "symmetrical center" of a symmetrical grouping of moving objects, one knows that it is in fact the "expansion center" or center of expansion (or of contraction).

*Expansion and Rotation*

Now imagine that all these people are walking on a large piece of metal plate which is rotating about a pivot (thus, even the center person is rotating). Or imagine that the plate is not rotating, but rather while these people are walking, they are all also rotating about the pivot, that is, the "center" person is not standing still; he is also rotating about the pivot (see Figure 8–4).

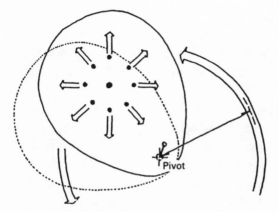

Figure 8–4

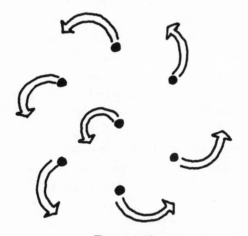

Figure 8–5

In this case we can say that there is a symmetrical expansion pattern which is rotating. However, the center of rotation (the pivot point) is not the same as the center of expansion (see Figure 8–5).

## The Geometrical Center

We now introduce one more slight complication. Assume we have a universe, which even when empty still exists. That is, the "space" of the universe is not a collection of "empty spaces between things" but is rather one big empty space. Whether or not this space can possess a "geometric center" depends on its fundamental nature. A universe can be finite or infinite, bounded (possessing an end) or unbounded.

If the universe is finite but unbounded, that is, it is closed (e.g., spherical), then (in analogy) to travel on the Earth's surface, one can continue going forever in any direction, but one finds that however one travels, one always returns to familiar places after a while. In such a case, any point is as much a geometric center as any other. If the universe is infinite (and unbounded), that is, it is open, (e.g., a plane), then from any point the distance to the end of the universe is infinite in all directions, so that one could say that at each point the universe is a circle with a radius of infinity which has that point at its center, or one could equally say that no point is the center. If the universe is finite and bounded (i.e., it has an "end"), then there is a unique point which is the geometric center. However, it is rather impossible for us to imagine such a strangely truncated universe. Similarly, nothing physical (that we can now conceive of) can be both infinite and bounded.[8]

### Center of Matter and Center of "Matter–Space"

In this empty universe we now place matter. If the amount of matter is finite and the universe is either finite and bounded or infinite and unbounded, then this matter has a center of gravity, and the area of space it occupies has a center—the "center of matter–space."

If the universe is finite but unbounded, it may or may not be possible to define such a (unique) center (in analogy to the surface of the Earth which has no unique center). (See Figure 8–6.)

### Center of Symmetry

As far as is known today, the universe is composed of galaxies which are composed of stars, some or all of which have satellites. These galaxies are grouped in clusters, and the clusters are grouped in superclusters, and the superclusters. . . .

Is this pattern symmetric? (See Figure 8–7.)

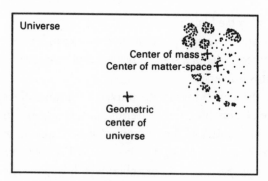

Figure 8–6

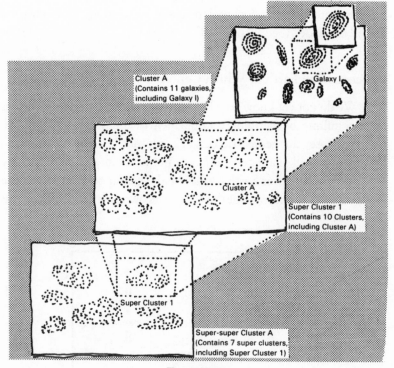

Figure 8–7

*The "Center" of the Universe*

In this space we now put our rotating, expanding (symmetrical?) system of objects (people, stars, dust particles, whatever). What is the center of this universe?

Since the only things in the universe are these objects, the center of these objects (i.e., the center of matter–space) can perhaps be considered the center of the universe. Or perhaps the true center is that of expansion or of rotation? Or perhaps it is the geometrical center of the total space (if it is definable) which is really the center? Or perhaps the symmetry center?

Thus, there are now five types of centers (the geometrical center of the universe, the center of matter–space, the center of symmetry, the center of expansion, and the center of rotation), and they need not all coincide! (See Figure 8–8.)

## Geocentricity: Earth at Which Center?

We do not yet know conclusively whether or not the universe is finite or infinite, but certainly it would seem that it must be unbounded. In either case, the geometric center could not be defined uniquely; however, we will examine whether or not it has any meaning per se.

It does seem that Earth cannot be the center of symmetry of the universe since it is located at the edge of its own galaxy. However, it is

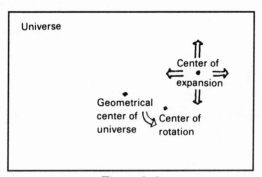

Figure 8–8

not at all clear that the universe does have a general spatially symmetric distribution of matter, so that the universe might well have no such symmetry for Earth to be the center of.

Since it is not known whether or not the universe is finite, and whether or not matter fills all the available space, the question of whether or not there exists a center of matter–space is also moot.

In any case, we shall discuss the possibility of a rotation center and an expansion center. In addition, we shall investigate the possibility of whether or not one can define at all a geometric center.

We will now examine the meaning of the geocentricity of our universe in the light of the preceding discussion. That is, we will examine modern cosmology to find out where, in our universe, might be these three types of centers. We will deal separately with each type of center, discussing the validity of geocentrism with respect to each type. Our first discussion will deal with the concept of the geometric center of the universe.

## Geocentricity and the Geometric Center of the Universe

*Introduction*

In ancient times it was believed that the universe was composed of two types of entities: space and matter. Space was the framework in which matter moved. Space was "absolute" in the sense that it was an actual entity, not simply the lack of matter. As an entity in its own right, it had an orientation, a center, and edges. Thus, any planet in the universe could be described according to where it was relative to the center of space, how it was moving relative to space (which was stationary), and what angle its rotational axis and orbital plane made relative to the inherent directions of space.

Thus, even if the universe had only one body in it, it would make sense to talk of its motions and so forth; there was no need of other bodies to provide a reference relative to which its motion could be defined. Space itself, absolute space, provided the reference. Motion meant motion with respect to absolute space.

In fact, it was believed that if there were nothing at all in the universe, "space" still had some meaning, and that this space would have a geometric center. Earth was believed to be located in the geometric center of space. However, when the concept of absolute space began to be discredited, and space ceased to be thought of as an entity in itself, speaking of a center of the universe began to seem ridiculous. Thus, it seemed that the universe could not possibly have the Earth at its center, since it had no center at all.

With the advent of modern cosmological models and the general theory of relativity, it was shown that space (time) is a "thing" in the sense that it can be bent and twisted! However, the controversy over whether or not absolute space exists, and thus whether or not a geometric center exists, is not quite over. We now deal with this question in detail.

### The Geometric Center of the Universe: Absolute Space

Until a few centuries ago it was believed that space was absolute. That is, space had meaning even if it was totally empty. "Space" was thus not merely the empty areas between pieces of matter; rather, it was a "thing in itself."

Thus, space could be considered to have a real, geometric center. However, for an entity to be a "thing" rather than a concept, it must have physical properties. Thus, for space to be considered as a "thing in itself" it had to have physical properties even when empty of all matter (energy). We now briefly discuss some of these supposed properties.

### The Ancient Belief that Space Is Absolute, Material, and Desires Rest

Foremost among the properties of space was that there was some "material" substance to it which enabled one to speak of speed relative to space, rotation relative to space, and position and direction relative to space. This "materiality" was later given the name "ether." For brevity, we will refer to these properties of absolute space as their *ether properties.*

In addition, matter was believed to "prefer" a state of rest, so that all moving bodies would eventually come to a stop without any outside interference. Thus, just as resting bodies need a push to get them to move, it was believed that they also come to an eventual stop on their own if the push is discontinued.[9] This property of matter, its "desire to remain at rest," was considered to be a result of some spiritual nature of space and matter that caused them to be "perfect" and thus to prefer "stability," "rest," and the like. We refer to this supposed property of absolute space desiring rest as its "perfection property."

Since the sun and planets rotate continuously about the Earth, they were considered either to have a continuous push on them given by God or angels or to be intelligent beings moving on their own volition. In fact, a circle was also believed to possess the attribute of "perfection" (it was the "perfect shape"), and the "perfection" of "heavenly" bodies was reflected in their "perfectly circular orbit."

### The Reasons for Belief in Absolute Space

These two absolute properties—of "ether" ( or materiality) and "perfection" (or stability)—were introduced in ancient times as a result of religious/metaphysical beliefs—not as a result of careful experimental observation.

In modern times, when explanations of the workings of nature began to be based on observation rather than on purely speculative philosophical "wishful thinking," these properties came under attack. According to the principles of the scientific method, if a concept is not necessary to the explanation of natural phenomena, and its ramifications are also not observed in nature, then it is assumed to be a nonvalid concept.

In our case, if the concept of absolute space could be shown not to be necessary to the explanation of phenomena, and if no manifestation of absolute properties of space can be observed, then it could be assumed that there was no such "thing" as absolute space.

We now examine the concepts related to the existence of absolute space in the light of modern physics.

### Discrediting the "Ether Property"

Earth revolves about the sun. The sun rotates about the center of the galaxy, the galaxy is receding from the other galaxies. . . . Everything seems to be in motion. Is anything standing still?! Of course, we feel ourselves to be standing still, but someone on the moon sees us as moving, and someone on Mars sees both of us moving and someone on Venus. . . . Thus we must rephrase our question to be: Is there anything in the entire universe that everyone everywhere must agree is not moving? A little thought will show that this would seem impossible. Of course everyone everywhere can *agree* to consider one particular something as standing absolutely still, even though this means that they all must agree that everyone else (including themselves) is moving. However, they could all just as well agree to consider some other one thing as being still. The choice of *which* particular one thing will be chosen as the agreed-upon stationary object can be a totally arbitrary choice. *Any* object whatsoever can be chosen to fulfill this role. This chosen object will then be, *by definition*, stationary. However, since each person can claim that his planet, rocket, or spacesuit is stationary, then there is obviously no thing which all *must* agree is stationary.

Thus all motion is relative. That is, since each can say he is stationary (and there exists no unique stationary point in the universe to contradict this), a statement about being in motion is meaningful *only* if one specifies the particular entity relative to which one is in motion. Thus, it would seem that the concept of an absolute space which possessed ether properties was quite unnecessary and probably invalid.

*Discrediting the "Perfection Property"*

Changing one's speed or direction of motion, or both, is called "acceleration." Acceleration brings along with it pushes and pulls, i.e., "forces"—a force pushing back when one accelerates forward, a force sideways when (accelerating by) moving in a circle. (This will be explained in more detail in the section "The Rotation Center.") What is the cause of this accelerator force?

The answer is that it is not really a force; rather, it is inertia manifesting itself as a "force." Newton's law of inertia states that objects at rest remain at rest, unless pushed or pulled, and objects in unaccelerated motion remain in unaccelerated motion unless pushed or pulled.[10]

This, of course, is in opposition to the perfection principle, which (erroneously, as we have seen) demands that all objects tend to a state of rest.[11]

Thus, the principles of both the "ether property" and the "perfection property" have been discredited. Since these were the properties which gave to space its "thing-in-itself"quality, does the invalidation of these properties imply that space is not a "thing"? In other words, have we disproved the existence of "absolute space"?

*Negation of Absolute Space*

Although the "perfection property" has been discredited, one could say that a new one—the laws of "inertia"—has been substituted for it and that there was therefore no damage done to the belief in absolute space. That is, empty space possesses a property; it induces inertia in matter (when matter is placed in it). Thus, having a property, space is a "thing." As a "thing," space is "absolute" and can have a center.

This is not necessarily the case, however, for Ernst Mach (a philosopher scientist whose work greatly influenced Einstein) claimed that inertia was not a property of the universe but was rather a property of matter itself. That is, inertia is a property of matter which causes it to interact with other matter (in space). Thus, again, space was left with no properties of its own and could not be considered "absolute."

In order to understand this issue and to see whether or not one can indeed attribute inertia solely to matter, and not to space, one must understand inertia better. For example, although we now know the properties of inertia, we still do not know its origin, i.e., what causes an object to have inertia.

## INERTIA

To help in partially answering this, we proceed as follows: Imagine the universe to be totally empty except for one "planet." This planet is accelerating. Now, how do we know that it is accelerating? As we saw before, motion is meaningful only if it is motion with respect to something. Since the rest of the universe is empty, there does not exist anything else relative to which the planet is moving! There is thus no way one can give any meaning to the statement that the planet is moving. Or is there perhaps? Since it is accelerating, inertial forces will be manifested and will serve as proof of its motion. (According to the equivalence principle, inertial forces can be attributed to the effect of unseen gravitational sources. In accelerating-up elevators, the floor presses against us as though gravity has increased! Thus, inertial forces do not unequivocally indicate acceleration, and the motion could be denied. However, in our "almost empty universe" there can be no hidden source. We must admit that it is experiencing a force because it is accelerating! So that even in an empty universe, the very existence of this force would unequivocally indicate that the planet was accelerating.)

Thus, there *would* seem to be meaning to acceleration in an empty universe.[12] However, it seems almost impossible that in an empty featureless universe one can notice any type of motion. It is quite illogical. Yet, the presence of the acceleration force, the inertia, seems to say otherwise. Thus, it would seem that the presence of the acceleration force (i.e., inertia) points to the existence of absolute space!

The resolution? We follow our logic and postulate that not only is there no inertia in a *totally* empty universe, but a planet in an otherwise empty universe also has no inertia. (We neglect self-interaction of particles and intra-object interaction of constituent particles.[13]) Thus, there is *no* meaning to the statement that the planet is moving.[14]

Of course we cannot merely create convenient postulates arbitrarily. They must be consistent with the rest of physics. Thus, we must ask how could it be that there is no inertia in an "almost empty" universe? Our question is based on the observation that in the universe (as we see it) there *is* inertia; it would therefore seem that one could extrapolate to the case of an almost empty universe and deduce that it would also manifest inertia. However, this may well be an invalid deduction, that is, we reason as follows:

If the universe was *not* empty, why *would* a planet have inertia? The answer: Every piece of matter in the universe creates a "field," and it is the cumulative field which is the source of inertia of every piece of matter which encounters the field. (The more mass in the universe, the more inertia.) Since this inertia field permeates the entire universe, all matter in the universe has inertia. However, if there is only *one* object in the universe, it has no inertia because there are no other objects in the universe to give it inertia.[15] Thus, there are no inertial effects of accelerated motion for such an orphan planet. We must therefore conclude that there is no way of detecting any motion in a universe with only one object. Indeed, in such circumstances, motion is actually undefined.[16]

We can now see the significance of this answer to our question of the existence of absolute space. Since there is no way to detect any motion, even accelerated motion, in an "otherwise empty" universe, then the "ether properties" of space cannot be observed. Since the "ether property" is an unnecessary concept to physics, it is invalid scientifically. Therefore space cannot be said to be absolute in that sense. In addition, since inertia is a property of matter as it interacts with other matter in space, rather than being a property of space *itself* (there is no "inertia" in an empty universe), the empty universe does not possess any "perfection property." Thus, it would seem that empty space possesses no physically relevant properties and, thus, cannot be said to exist!

This was not the last word, however.

## ABSOLUTE SPACE REVISITED

In 1949 and 1969, solutions of Einstein's equations were found which seemed to imply that absolute space did indeed exist.[17] One solution corresponded to a rotating universe empty of all matter–energy. Now, as we have seen, rotation, as all motion, should have no meaning even in an almost empty universe. Yet here the universe was totally empty, and yet the whole universe was rotating. Very strange! Further research showed that perhaps there was indeed some energy in this universe and, therefore, perhaps it was not really empty.[18]

According to the principle laid down by Ernst Mach, inertia could exist only when matter–energy existed, and empty universes could not rotate since they had no body relative to which they rotated. The controversy over the possible existence of absolute space is thus actually a controversy over the validity of Mach's principle. This controversy has still not been resolved.[19]

Until the existence or nonexistence of absolute space has been established (and, thus, also the existence or nonexistence of a unique geometrical center), it is not possible to anticipate the possible questions and solutions this decision will generate.

It is moot whether or not the concept of a geometrical center is meaningful and whether or not the Earth could be considered to be located at such a center even if it *did* exist. We thus put aside the geometric center and consider geocentrism only in relation to the expansion and rotation centers of the universe.

## THE GENERAL THEORY OF RELATIVITY: EXPANSION AND ROTATION CENTERS

The gravitational force at a particular place, due to the presence of mass–energy, manifests itself as a curvature of space–time at that place. If there are not nongravitational forces acting on it, any entity in that location will move on a geodesic of that curved space (something like the "shortest possible path"). In the absence of gravitational sources (matter or energy), space–time is flat and the geodesics are ordinary straight lines. Therefore, particles not acted on by gravity (or other forces) move in a straight line with a constant speed and direction. (Standing still represents the constant speed zero.)

A changing pattern of mass–energy gives rise to a changing spatial curvature. Space is therefore neither absolute nor static. It is dynamic.[20] One of the most interesting examples of this fact is explored in the next section.

## THE EXPANDING UNIVERSE

The universe is expanding!

This extraordinary phenomenon was deduced from the observations of astronomers and can be deduced from one of the solutions of

Einstein's equations. When one says that the universe is expanding, what is meant is that the actual "fabric" of the universe (i.e., space–time) is expanding.

How can space, let alone space–time, expand? Just as an aid in imagining this, let us visualize a balloon with dots painted on it so that all dots are equally distant from all their closest neighbors[21] (see Figure 8–9).

Now the balloon is blown up more. As it is inflated, the dots move further apart (see Figure 8–10).

Now if a flea were to stand on a dot and watch the neighboring dots, he would see them all moving away from him as the balloon expands. In fact, whichever dot is chosen as the viewing point, the viewer will see all the other dots moving away from him. Thus, if each dot had an observer on it, each would see everyone else moving away; since each one sees all the dots receding from his viewing point, each would assume that he is at the center of the pattern. Thus, all assume this,

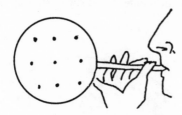

Figure 8–9

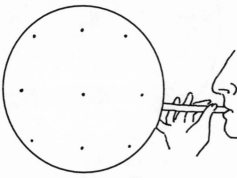

Figure 8–10

and all are correct. Every point is the center of expansion! (see Figure 8–11).

## BACK TO SPACE–TIME EXPANSION

Now if space–time is expanding in a manner analogous to the balloon, at all points in the universe the expansion seems to be centered right there![22] (see Figure 8–12).

Each dot represents a galaxy (some of which are numbered). The recession pattern of the galaxies as seen by any observer (e.g., number 3) is identical to the pattern seen by any other observer (e.g., number 4). Each observer sees the galaxies recede from him with speeds proportional to their distances from him. The farther away the galaxy is, the faster it recedes. Thus, any observer in the universe, on considering the expansion of the universe, could legitimately conclude that the "center of the expansion" of the 3-D universe is precisely his location, i.e., he is the center of the universe!

Indeed, observations by Edwin Hubble showed that this is actually the case for our universe. He showed that all the distant galaxies

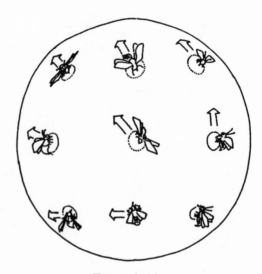

Figure 8–11

**Figure 8–12.** Each dot represents a galaxy. (Some of them are numbered.) The recession pattern of the galaxies as seen by any observer (e.g., number 3) is identical to the pattern seen by any other observer (e.g., number 4). Each observer sees the galaxies recede from him with speeds proportional to their distances from him. The farther away the galaxy is, the faster it recedes.

are receding from us, and that the more distant the galaxies are from the Earth, the faster they are receding from us.

There are two possible explanations for this: Either our galaxy is at the unique center of a static universe, and all the others are receding from this center, or this pattern of recession would be observable at any place in the universe and is due to the expansion of the very space–time of the universe itself.

Scientists generally assume that the Earth is not located in a special part of the universe, and that it is not uniquely related to the observed recession of the galaxies. It is assumed that whatever pattern is viewed from the Earth, it would be seen elsewhere as well. That is, it is *assumed* that our galaxy is not the center of the universe. With this assumption, the observed pattern of galactic recession—centered about us—implies that the universe is expanding.

Thus, even under the assumption that the Earth is not the unique center of the universe, the Earth is nevertheless a relative center of the pattern of recession of the universe.

Thus the pendulum has swung back, and as far as the expansion of the universe is concerned, modern science has reintroduced the validity of the geocentric view, albeit in a very sophisticated and relative sense.

We have seen that the Earth can be considered as a center of expansion of the universe. Even more important, however, is the question of whether or not the Earth can be considered to be at the rotational center of the universe; the reassignment of the rotational center from the Earth to the sun was the essence of the Copernican Revolution, which started the whole seeming conflict between religion and science in modern times.

## THE ROTATIONAL CENTER OF THE UNIVERSE: UNIFORM MOTION VERSUS ACCELERATED MOTION

Probably everyone who reads this has at some time traveled through the air at speeds of several hundred miles per hour. Those who have had smooth flights know that it is possible to gently sip a drink without spilling a drop, that a walk in the aisles needs no special skills; in fact, during the greater part of the flight it is almost impossible to detect any effect of the motion (without peeking out the window!). Similarly, during a smooth car ride, riding (even at high speed) can be indistinguishable from the wait at a traffic light. It is only while the car is speeding up or slowing down, or turning towards the left or right, that some effect is noticeable. Similarly, in a plane, there is an effect only when the plane takes off, lands, or banks.

If a car speeds up, we are pressed back in our seat. If it slows down, we are pushed from the back of the seat. If it turns right, we slide to the left; if it turns to the left, we slide to the right. (If a plane, or elevator, rises suddenly, we are pressed to the floor; if it descends suddenly, we are left somewhat suspended off the floor.) Thus, it is only change of speed and change of direction which lead to noticeable effect. Motion with both constant speed and constant direction is indistinguishable from staying at rest. Motion with constant speed but a change of direction, or constant direction but changing speed, does lead to noticeable effect.

In order to have a concise terminology, we make use of the term *acceleration*. This means a change in the state of motion—either in the direction of the motion[23] or in the speed of the motion,[24] or in both simultaneously. Nonaccelerated motion means motion with both a constant direction and a constant speed  (see Figure 8–13).

Now we can summarize all of the above in a statement which is part of, and follows from, the fundamental principles of Einstein's theory of special relativity: There is no way for anyone who is moving without acceleration to determine whether or not he is actually moving, unless he peeks "outside." (The principle of relativity contains more than the statement given here.)

## RELATIVITY OF UNIFORM MOTION

Now, suppose a physicist is in unaccelerated motion[25] and believes himself in fact not to be in motion at all. He performs a physics

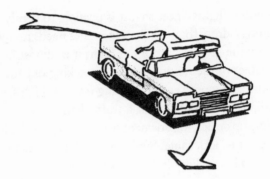

**Figure 8–13.** The path is toward the "outside" of the curve.

experiment. Of course, since there is no way for him to determine that he is in motion, the results of the experiment must be indistinguishable from the results he would obtain were he in fact to be stationary. He then "peeks" and sees the landscape flashing by him. To him, it seems as though he is standing still, and all the world is rushing by. (This is a very common illusion on trains.) Outside stands another physicist who performs the identical experiment. He of course also obtains results consistent with his being at rest, but these results are then necessarily the same as those of the physicist on the speeding train.

Now, the physicist on the train feels himself to be at rest.[26] In addition, his experiment verifies that he is at rest, for his results are identical to those of an experiment in a resting train. He can explain the fact that the whole world seems to be flashing by him by simply claiming that indeed the train *is* standing still, and indeed the world is flashing by! Of course the physicist on the ground claims exactly the opposite. Although it would seem that the latter is correct, the principle of relativity states that *both* viewpoints are equally valid scientifically! If one person is moving away from another without acceleration, he may claim that *he* is actually remaining at rest, and that it is the *other* who is moving. Indeed, *both* can make the identical claim, and *both* points of view are equally valid!

Of course, if one of them were pushed back in his seat or caused to slide over, we could point him out and say: "Obviously, *you* are the one who is moving, and it is the other who is at rest." However, if both move without acceleration, this motion will be without sliding and without pushes, and therefore both can claim to be at rest. Tentative conclusion: *Uniform* motion is relative. *Accelerated* motion is not relative because of the accompanying effects.

And now we can even extend this principle. Even with accelerated motion, the principle of relativity is valid! How is this possible?

## EINSTEIN'S ELEVATOR

We can feel the effect of the Earth's gravity on our body when standing by the feeling of our feet pressing against the floor or of the floor pressing against our feet, which is the same thing. Essentially the same type of sensation results from accelerations (changes of speed or direction).

In an accelerating or decelerating car (speeding up, slowing down, or turning), we are pushed back into the seat or thrown to the side. Equivalently, we can say that we feel pressures on our backs from the seat and sides of the car. These are essentially the same types of sensation as induced by the pressure caused by the gravitational pull of the Earth—pressure on our feet when standing, on our legs and back when lying down, or on our backs and sides when in an accelerating car. Indeed, physically, they are indistinguishable.

In express elevators in skyscrapers, the acceleration of the elevators is greater so that an elevator can reach a high enough speed to make the long trip in a reasonable amount of time. One can feel the pressure of the floor pressing against one's feet, more so than usual, and a reverse sensation (of lesser pressure than usual, often felt in the stomach) as the elevator slows down. (Riding any ascending elevator, one can feel all these sensations, but to a lesser degree.) When accelerating on the way up, we feel as though we weigh more—or as if the Earth became suddenly more massive, with a concomitantly greater gravitational attraction. In the same way, we feel as if we weigh less when decelerating on the way up. (In a descending elevator, the order of sensations is reversed.)

Einstein demonstrated that at any given small area, it is, in effect, impossible to determine by experiment whether a given "force" arises due to a gravitational pull of some mass or whether the force is a result of acceleration.

If the area being studied is large enough, however, then changes in the force over the extent of the area being studied will reveal its nature. If it is an acceleration, it will not change; if it is a gravitational effect which is due to a nonuniform field, it will vary. (On the Earth, for example, if the gravitational field varies enough over the body of a person for the field to be detected by very sensitive instruments, we would know that the force is due to gravitation rather than to some acceleration.)

Thus, in theory, when examining a force which is uniform over a certain extent, it is impossible to determine whether the force is the result of a uniform field with that extent or is due to an acceleration.

## RELATIVITY OF "CIRCULAR" ACCELERATION

Anyone who has ridden a merry-go-round knows that there is a force pulling away from the center during the ride. This force is actually due to the accelerated motion (even if the merry-go-round is revolving with constant [angular] speed, one is constantly changing direction) (see Figure 8–14) and is the same force that causes people (in turning cars) to slide over in their seats when the car they are riding in is moving along a curve.

The person on the carousel, however, need not admit *he* is undergoing acceleration (changing direction), even though the world is spinning around him and he is being pushed to the outside. *He can claim that he is stationary* and the *universe* is spinning and that the force pushing him to the side is due to the gravitational "rotation force" of the spinning universe! The general theory of relativity says that this may be a valid viewpoint! (see Figures 8–15 and 8–16).

A number of problems seem to arise involving this viewpoint; these problems, and an outline of their solutions, are presented in the following:

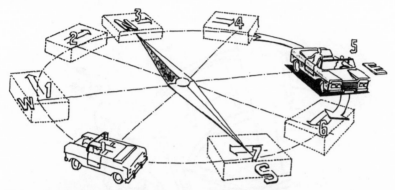

**Figure 8–14.** The car is moving about the circle with constant speed. The direction of the motion of the car is, however, not constant—it is continually changing. E.g., at position 1, it is headed due North; at position 2, it is headed North-East; at position 7, it is headed due West. Thus motion in a circle always involves a change of direction (and thus, by definition, it involves acceleration), even when the speed is constant. (One can say that he is changing his direction in a constant way, but one cannot say that his direction is constant!) The combination of speed and direction is called "velocity." Thus, in this case, the car has a constant speed, but because its direction is not constant, its velocity is not constant.

Imagine someone spinning on his heels (e.g., one revolution per second) under starry skies. According to the interpretation of relativity we have discussed, one could consider that the person is stationary and the entire universe is spinning about him. This, however, gives rise to a number of questions:

1. The universe rotates once per second, and the circumferential distance traveled by each star is fantastically huge, so that the speed of each star must exceed (by far) the speed of light—which is forbidden by relativity theory itself.
2. As soon as the person begins to spin, one can say that the universe is spinning instead. However, if the universe begins rotating *immediately*, then the effect of his motion must have traveled to the stars instantaneously, i.e., faster than light.[27] This is impossible according to relativity theory.

**Figure 8–15**

3. The universe rotates only because the person moved his feet. Yet how could he have enough energy to cause the entire universe to move?

4. Is it not true that the "centrifugal force" caused by the motion of the stars is a "fictitious" force? Thus, does this not prove that one cannot *really* consider the universe to be in motion?

## ANSWERS

In the Machian special relativistic interpretation, the inertia of the Earth is due to inertial fields in its neighborhood—those fields that have had time to reach there from their sources (the stars and so forth). Each component of the inertial field of the Earth is therefore due to the configuration of matter–energy as it was when that component began its journey to Earth—at the speed of light.

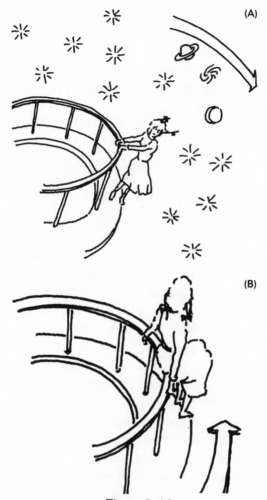

**Figure 8–16.**

(A) Whenever one feels a centrifugal force, one also notices the rest of the universe spinning around oneself. Mach stated that this is not only a correlation, it is a cause–effect. That is, it is the effect of the universe rotating about us, which we feel as a centrifugal force! Thus the girl on the merry-go-round can attribute the force she feels on herself to the rotation of the universe about her and considers herself to be stationary!

(B) If there was nothing else in the universe but the girl and the merry-go-round, the girl of course would not see anything spinning about her. Mach claimed that she would not feel any centrifugal force and would have no inertia! The fluid in her ears would not move, so that she would have absolutely no sensation of motion. Of course, defining motion at all in this case is probably impossible.

Thus, when the Earth rotates relative to the stars, the centrifugal force on the Earth can be explained as being due to the relative rotation between the Earth and the inertial field near the Earth—near the Earth but originating from far away.

The effect does not depend on what is "really" moving, only on the existence of relative motion. Also, the effect is not between the distant stars and the Earth, but rather between the local inertial field of the distant stars, and the Earth. Therefore, there are no problems of causality involved. The answers to our questions above then are:

1. Relativity theory forbids the transfer of matter–energy or of information at speeds faster than that of light. However, for there to be a true violation of this law, the matter–energy must move faster than light in a "local" frame.[28] In our case, it is only in the frame of the spinning person that the stars are moving faster than light.

2. On consideration, one can see that no information can be transmitted via this method, i.e., someone residing on the far stars cannot instantaneously receive any messages from the spinning person and vice versa. (Even if the spinner starts and stops his motion in a sort of Morse code, this will not transfer any information[29] since the star resident does not feel any rotation; he knows of the spinner only when he finally sees him spinning many years later when the light reaches him.) Similarly, no information is transmitted by the spinner's seeing the stars begin to rotate instantaneously.

3. Let us look at the spinner before he begins his motion. He is at rest relative to the universe. Imagine that the entire universe is rotating with him, but that since everything is rotating in the same way, it is not noticeable. Then the person moves his feet to spin. By doing so, he is introducing a relative rotation between himself and the rest of the universe. Since the rotation was unnoticeable previously, one can choose the direction of that rotation arbitrarily (in each reference frame). One can then say that the person is now rotating against the previous motion of the universe-as-a-whole. Thus when he says that the universe is now rotating, he has not contributed the necessary energy to accomplish this! Rather, the universe was rotating before he moved; he simply notices it now because of his motion.

4. Physically, the centrifugal (and Coriolis) force can be measured just as can any "real" force. It is only "fictitious" if we do not consider the universe to be rotating! (If we *do* consider the universe to be rotating, we need not answer this question since it doesn't even arise. The question is valid only if one assumes that the universe does *not* move; thus it can certainly not be a proof against the motion of the universe.)

Dynamical laws as expressed in Earth rest-frame coordinates are asymmetrical and cumbersome, but even in classical physics they are nevertheless valid expressions of the laws of dynamics. That is, they are complicated equations which give the same correct answer as the simple equations resulting from laws formulated in the rest frame of the sun. What general relativity adds to this is the fact that there is no *physical* means of distinguishing one frame from the other. There is no absolute space, and therefore there is only relative motion.

There can, however, exist a preferred frame—preferred physically, not only mathematically. For example, the frame in which a Foucault pendulum does not rotate and the frame in which the cosmic background radiation hitting the Earth is the same in all directions. This preferred frame exists because the particular solution to the relevant cosmological equations which our universe has picked out is not symmetric even though the laws of physics themselves are symmetric. (This is a typical example of *symmetry breaking*.)

## GEOCENTRICITY AS A RESULT OF RELATIVITY

We return to our original problem, that of the geocentricity of the universe.

From the general theory of relativity we can see that one can say that the Earth is at rest, and the sun—in fact the entire universe—is revolving about it![30] The centrifugal force which has been measured and attributed to our circular motion about the sun can in fact be attributed to the gravitational "rotation effect" of the rotation of the universe about us!

Thus, the Earth can claim to be the "center of rotation" of the universe! Of course, *any* point in the universe can claim to be the center of rotation of the universe, as long as it does not claim to be the unique center. Thus, nonabsolute geocentrism is just as valid as a nonabsolute Copernican system.

The theory of relativity does not say that geocentrism is *correct*; rather, it denies the *absolute* significance of *either* theory.

The importance of the Copernican revolution, however, was that it denied *any* validity to the geocentric views of the prevalent cosmology at that time (i.e., the Ptolemaic system) and claimed truth for itself only; thus, as the philosopher of science Hans Reichenbach wrote:

> The relativity theory of dynamics is not a purely academic matter, for it upsets the Copernican world view. It is meaningless to speak of a difference in truth claims of the theories of Copernicus and Ptolemy; the two conceptions are equivalent descriptions. What had been considered the greatest discovery of western science since antiquity is now denied its claim to truth.[31]

Relativistic physics is entirely concerned with relations between events, not the private perceptions of these events. The laws of phenomena must be the same whether they are described as they appear to one observer or to another. The general coordinate-free formulation of relations between bodies describes the interaction between them in terms of relative accelerations, not in terms of the motion of this or that particular body.

When this is translated into the coordinates of a particular frame, we obtain a visual geometric pattern of the motion as seen from that frame. These patterns are simpler in some coordinate systems than in others. Similarly, the equations, when expressed in terms of some particular coordinate system, may be simpler than when expressed in terms of some other coordinate system.

In our case of the relative motion of the Earth and the sun, the equations describing the relative motion of the sun, Earth, and planets are arrived at by solving certain gravitational equations. These equations—and the visual pattern of motion they describe—are sim-

Avi Rabinowitz 117

pler in some coordinate systems than in others. For example, in a co-ordinate system (reference frame) based on the sun as nonmoving, all the planetary motion is elliptical about the sun—a very simple pattern; whereas in a coordinate system (reference frame) based on the Earth as nonmoving, the visual pattern of the planetary motion is quite complicated, as are the equations describing that motion.

Therefore, even if we speak of the sun moving instead of the Earth, the required relative motion between the Earth and the geodesic of space–time is still taking place as demanded by the gravitational equations. It will simply seem to be a more complicated motion when viewed in this way.

One cannot "explain" why the planets move as they do other than to solve the equations and find the answer. The solution to the equations is, then, the "explanation." To explain the motion of the planets and the sun as seen from Earth, one must simply solve the gravitational equations for the Earth–planets–sun interactions in the reference frame of the Earth. The resulting complicated equations will describe exactly the complicated geometrical motions of the planets and sun as seen from the Earth.

This is similarly true for the motion of the stars as seen from the Earth. The law of gravitation does not say who moves. It only dictates what the relative motion will be like as seen in any reference frame.

Of course, a description of the universe in terms of a non-geocentric system is simpler.[32] However, in Reichenbach's words:

> the idea of simplicity cannot be used to decide between the Ptolemaic and Copernican conceptions. The Copernican conception is indeed simpler, but this does not make it any "truer," since this simplicity is descriptive. The simplicity is due to the fact that one of the conceptions employs more expedient definitions. But the objective state of affairs is independent of the choice of definitions; this choice can result in a simpler description, but it cannot yield a "truer" picture of the world. That these definitions, e.g., the definition of rest according to Copernicus, lead to a sim-

pler description, of course expresses a feature of reality and is therefore an objective statement. The choice of the simplest description is thus possible only with the advance of knowledge and can in general be carried through only within certain limits. One description may be simplest for some phenomena while a different description may be simplest for others; but no simplest description is distinguished from other descriptions with regard to truth. The concept of truth does not apply here, since we are dealing with definitions.[33]

Of course, this does not necessarily mean that those who believed the universe to be Earth centered did so out of an understanding of general relativity. Nor does it mean that we should necessarily interpret the Torah literally when it speaks of cosmology.

It is interesting, however, that expressions in the Torah which seem to imply a geocentric cosmology can still be understood literally. Moreover, these expressions implied beliefs that during the period between Copernicus and Einstein were considered scientifically false. This was a time when much of the Bible was allegedly disproven. Nevertheless, the Torah was vindicated.

There may well be other difficult periods ahead when truth is sacrificed to preserve validity—yet later reinstated with the emergence of new ideas.

## PART THREE: THE SIGNIFICANCE OF MAN

Until the time of Copernicus the picture of the universe most people carried in their minds was that of a "small" universe. It was natural to consider the Earth to be the most important body in the universe since the other bodies appeared to rotate about it and serve it: the sun for light during the day, the moon for light at night, the planets and stars as celestial lights whose patterns foretold future Earthly events. Man was naturally accorded the status of sole intelligent being in the universe and was sure of his significance and destiny.

Then came the Copernican revolution, which plucked man from the center of the universe and relegated his home, Earth, to the status of merely one planet among many. Giordano Bruno (1548–1600), basing himself on the Copernican system, preached the infinity of the universe[34] and the infinite multiplicity of life forms in it. Man's status seemed to change from sole intelligence in the universe, residing at its center, to simply one of possibly infinitely many races of intelligent beings, occupying a non-unique position in one of infinitely many galaxies. To many, the new astronomy seemed to indicate that man was simply insignificant. This of course could not be *proven*. However, it was generally felt that the belief in man's significance was simply incompatible with the idea of an infinite universe.[35]

## THE JEWISH VIEW

The universe was created by God as a vehicle through which some great Purpose would be achieved. Each individual element of the universe was specially designed to further the goal of accomplishing that Purpose. Indeed, after each stage in the Creation, God inspected His work and proceeded only when He "saw that it was good." Man, as a free-willed intelligent being capable of distinguishing between good and evil, has an important role in achieving the cosmic Purpose. Although he is a "creature of the dust" and returns to dust, his innate potential to rise above his animal nature grants him a status far above that of the dust he is composed of. The psalmist expresses this dual status poetically: "What is man that Thou shouldst remember him. . . . Yet Thou hast made him but slightly lower than the angels. . . ."[36]

However, man is only one part of the universe; all components in the cosmos have their purpose, and in some sense some of these entities can be considered to have a purpose even higher than that of man. Rambam wrote in *The Guide for the Perplexed*: "Most of the difficulties which lead to confusion in the search for the purpose of the universe arise from man's erroneous idea of himself and his supposing that all of existence is for his sake alone. An ignorant man believes that all of existence is for his sake . . . but if man examines the universe and understands it, he comprehends what a small part of it he is."[37] "The truth is that all mankind . . . are very low in comparison [with some other creations]. . . ."[38]

Nevertheless, the individual human being has an intrinsic worth which is far too great to sacrifice "for the good of the many," a sacrifice which is glorified in many societies. Indeed, Judaism teaches that each and every human being is given unique capabilities and can fulfill a unique role in aiding mankind in achieving its spiritual Purpose. For this reason, the rabbis taught that "He who saves one life is considered as if he saves a whole world."[39] Each individual is as valuable as all of humanity together!

Thus, in Judaism the importance of man, his significance in the universe, does not derive from numerical consideration; even in the modern world of megalopolises containing millions of alienated insignificant-feeling inhabitants, each individual is nevertheless a unique and irreplaceable element of the cosmic design.

Indeed, if there exist intelligent species other than man in this or another universe, one can be sure that they too have a unique role to play.[40] What of the role of man specifically?

Part of the Purpose of man was revealed by God via the Torah. Among other things, man is commanded to "walk in the ways of God" by emulating God's kindness, charity, mercy, and justice as well as His creative ability. All but the last quality seem obvious goals for man, but the last is somewhat enigmatic; after all, God creates ex nihilo while man can shape only that which is already created. However, in Genesis God commands man to spread his dominion over the world. Jewish tradition tells us that this means that man is to use his innate creative abilities to bring order to chaos. This man accomplishes by determining the laws of nature,[41] by building complex physical artifacts, by transforming raw materials into works of art, and most importantly by using his physical environment (including his body) to elevate the mundane to the level of the holy. In order to accomplish his task, man was given the ability and duty to achieve dominion over the lower living species and over inanimate matter.

Indeed, even now, at the very infancy of scientific endeavor, without stirring from our planet, without much expenditure of muscle power, we have, using our minds, discovered equations governing the nature of matter, the expansion of the universe, the collapse of stars into black holes, and so forth.

And not for long will we be constrained to do all this only from a distance. In only a few decades, man has learned to fly and to reach the planets. It seems inevitable that just as man has walked on the moon and sent robots to Mars, man could walk on the planets of many stars in many galaxies and "extend his dominion" over many worlds.[42] Thus, man's significance is indeed cosmic; his actions have the potential to affect, hopefully for the good, the entire universe. Indeed, it is his duty to do so, in order to help liberate the "spark of holiness" hidden in each part and aspect of Creation.

We now summarize the main "facts" causing overly existential Western man to believe so firmly in his own utter insignificance. Man is so small and weak compared to the rest of the universe; the universe is so vast that no one would even notice our disappearance. The Earth is simply a run-of-the-mill planet with nothing special or unique about it, and so there is nothing special about its inhabitants. The universe probably contains so many beings that man is irrelevant.

It is quite obvious that from the Jewish perspective, none of these "facts" are true. Indeed, they indicate a total lack of understanding of the meaning of "significance." Clearly, the fact of being large, or strong, or even unique, does not make something "significant" in the sense of having a "Purpose." In addition, if something *is* significant, it need not be *unique*; adding more of it does not reduce its intrinsic significance. Cosmic significance is not a quantity determined in the stock exchange according to laws of supply and demand; rarity and abundance are irrelevant quantitative concepts.

Because the Western concept of the significance of man, however, was based on the inherited pagan ideas of the physical prowess of man and on his egocentric geocentrism rather than on his moral qualities, when man's ideas of his strength and spatial location were challenged, so was his feeling of significance.

Opposed to this emotional, pagan response of Western man to the awesome beauty and grandeur of the cosmos stands logic and Jewish thought; it is in the light of these that we now analyze the question of man's significance.

## MAN IS SMALL AND WEAK
## COMPARED TO THE REST OF THE UNIVERSE

Man is small. In a religious sense, man is cosmically significant not due to physical size or strength but rather because he possesses a free will and the ability to distinguish between good and evil. Certainly a farmer gazing at his plowed fields does not feel insignificant, even though the fields are vastly larger in area than he himself is.[43] Certainly a king of millions of subjects does not feel insignificant when viewing the crowds greeting him. Thus, man's significance has nothing to do with relative size. Rather, it is a measure of a man's feeling of power or helplessness. If man has power over a large empire, he is not intimidated by its size.

Considering man's fantastic ability to scientifically uncover the secrets of the farthest reaches of the vast universe and to create suns and destroy planets and so on, there is no reason for man to feel overwhelmed by the size and power of the universe. Intellectually and physically, man seems to have the potential to totally dominate the inanimate universe. Man may be physically insignificant to the universe right now, but indirectly, through his progeny, man is potentially—subject to the Will of God—of supreme significance to the physical universe. Intellectually and physically it is the inanimate universe which should feel insignificant next to man, rather than vice versa.

## "NO ONE WOULD NOTICE
## OUR DISAPPEARANCE"

It is quite clear that one cannot judge relative importance in spiritual matters using the criterion of physical size alone. Certainly it is possible to imagine that one human being may be infinitely more significant spiritually than a large dead planet, a huge fiery star empty of life, or even an entire uninhabited galaxy. Thus, the fact that the Earth is only a tiny speck in the vastness of the universe is in itself certainly no reflection on the possible spiritual significance of man.

## EARTH IS NOT UNIQUE—IT IS NOT THE UNIQUE CENTER OF THE UNIVERSE AS WAS ONCE THOUGHT

Certainly it is clear that the fact that the Earth might not be the unique center of the universe has no bearing per se on the spiritual importance of man and his actions. Man's importance does not derive from the physical position of his planet in space or from its uniqueness; rather, his importance derives from the fact that he is a free-willed intelligent being. Of course it is true that the ancient and medieval philosophers assumed a cosmology with Earth at the center, and that this scientific "fact" was accepted by most people, including religious leaders. However, the details of the geometry of the universe are not relevant to Jewish religious belief, and human importance is not dependent in any way on the centrality of the location of Earth.

## "THE EARTH IS A RUN-OF-THE-MILL PLANET" AND ALSO "THE UNIVERSE MIGHT CONTAIN SO MANY OTHER BEINGS"

There is no such thing as a run-of-the-mill significant entity. An entity which possesses significance does not lose any of it simply because there are many such entities. Remember "He who saves one person has saved a world," for each person is unique! A planet full of unique beings is unique, regardless of how many other unique planets there might be.

Certainly no one would suggest that because there are four billion people on Earth, each human being is insignificant! Similarly, even if there were many billions of intelligent species in the universe, this does not decrease the uniqueness or significance of any of them.

Each intelligent species (if indeed there are more than one) might well fit into the pattern of Creation in a unique manner. Each species, whether on Earth or on other planets, indeed each individual being, is worth intrinsically no less than the rest of Creation. Thus, the fact that the universe is exceedingly vast, and may even contain intelligent extraterrestrial life-forms, in no way indicates that man is cosmically insignificant.

Far from being a blow to religion, the new discoveries of cosmology, and of science in general, serve to aggrandize and elevate our concept of God. Knowledge of the true vastness of our universe aids in neither the proof nor the disproof of the existence of God—indeed, these are *both* probably impossible to ever achieve. Nevertheless, all who truly understand the wonders of nature and appreciate its beauty are affected alike by a profound sense of awe. To those who believe in God the Creator, this increased understanding of the vastness of the physical universe contributes as well to an increased appreciation of the Glory of God. God is elevated from the King of a clump of earth to the Master of the infinite reaches of the universe—from the Creator of beings derived from dust to the Designer of a fabulously complex system of biochemical organisms composed of almost magical physical particles.

Truly, contemplation of nature in its manifest wondrous beauty can lead the believer to a deeper sense of love and awe towards God.

From *B'Or Ha'Torah* 5E, 1986, expanded and updated. Illustrations by Michael Drewes, copyright © by *B'Or Ha'Torah*.

# 9

# Modern Physics
# and Jewish Mysticism

*Gedaliah Shaffer*

The reciprocal transformation of matter and energy is a major theme of both modern physics and Jewish mysticism. These thought-provoking parallels give us a holistic picture of ourselves as Jews participating in the greater universe.

## MYSTICISM–SCIENCE PARALLELS

The secularization of Western man's worldview during the course of the scientific revolution[1] of the past three centuries has engendered a profound dichotomy between man's religious/mystical beliefs and his intellectual/ scientific perspective. Medieval man was able to achieve some type of synthesis between faith and reason. In contrast, for contemporary man, these constituents of his mental life have become polarized, thus producing the seemingly irresoluble conflict of religion with science,[2] much to the detriment of religious belief in the modern world. In view of the protracted period of historical evolution culminating in this position, and due as well to the normal delay required to overcome human intellectual inertia, it is perhaps understandable that the dramatic discoveries in the physical sciences during this century have not as yet profoundly affected this aspect of modern man's basic *weltanschauung*. If, however, one explores the philosophical ramifications of some of these discoveries, one finds, surprisingly, a demonstrable reduction[3] in the gap between the religious and scientific perspectives. What emerges from this analysis is a scientific perception of the universe which has, to a great extent, converged on that of the traditional mystical viewpoint that is central to religious thought.

This realization is of course no longer novel. A number of recent books point out some of the parallels between modern physics and various Eastern mystical systems.[4] These systems tend to be rather syncretistic and hence somewhat contradictory, thereby depriving these striking parallels of a certain degree of conviction. Whereas, to the believing Jew these systems are repugnantly idolatrous and profane, further discussion of this problem must be left for another opportunity.

Now is the opportune time to begin a program to develop the parallels between these new scientific conceptions and the eternal truths of Judaism. The Jew believes the Torah is a consistent, coherent picture of the universe which gives man profound insights into the underlying nature of existence in addition to ethical and moral guidance. An analysis of the extent to which the various subfields of modern physics have rediscovered the truth and profundity of these insights would constitute the content of this program.

Of course, in a paper of the present scope it is impossible to do more than superficially indicate a few of the more notable points of contact without doing justice to either field. One can only hope to demonstrate the feasibility and value of the program described and present a tantalizing glimpse into some of the fascinating insights to be obtained.

## SCIENTIFIC DETERMINISM
## AND HUMAN FREE WILL

The Western scientific tradition culminating at the end of the nineteenth century was all inclusive and thoroughly deterministic.

Given a "snapshot" of the state of the universe at a particular instant of time, the classical physicist felt he could in principle predict with complete precision and absolute detail the evolution of the universe for all time to come. If, in addition, he accepted a mechanistic picture of the human mental process,[5] he precluded the possibility of human free will. Of course, the religious perspective, ever concerned with man's moral accountability for his actions, denies that any wholly mechanistic view of the human mind can be valid. Although the sensory inputs and motor outputs of the mind may be reducible to electrochemical reactions, at this point in time it is essentially an act of faith on the part of the scientist to extend this to the full complexities of the human decision-making process and the capability for abstract thought. This extension is rejected by the religious perspective which views the human mind as ultimately dependent upon a nondeterministic, suprarational spiritual force called human free will. The difficulty in identifying the locus of action of this nondeterministic component and explaining how it can interact with the purportedly deterministic workings of the human sensory-motor system has been responsible for much of the seeming incompatibility of science and religion since the time of Descartes.

Quantum mechanics, developed during the early decades of this century, has completely superseded Newtonian mechanics.[6] According to the tenets of the Copenhagen interpretation of quantum mechanics, our knowledge of even the inanimate elements of the world cannot be complete because nature itself operates in ways which are not fully deterministic. The most we can predict about any physical system is the probabilities of its evolving in a variety of ways. In fact, not only is it impossible to deterministically predict the evolution of a system in time, but according to Heisenberg's uncertainty principle it is even impossible to specify the precise instantaneous state of a system at a given time. Thus, the possibility of absolute knowledge of the universe promised by classical physics is reduced to the educated guess of quantum mechanics.

It is obvious that any modern picture of the human mind (even if mechanistic) based on a full appreciation of the ramifications of quantum mechanics will be far more congenial to the possibilities of a role played by human free will. If even the lifeless elements of the universe around us cannot be completely described scientifically, if the evolution of inanimate matter can no longer be definitively predicted, how can even the most mechanistic theory of the human brain hope to completely describe and predict human behavior? Clearly, if we refuse to be satisfied with the hopeless ignorance of our ultimate state suggested by quantum mechanics, we must accept some nondeterministic element which transcends the possibilities of human science as presently envisaged to explain the physical world around us. It is only a small logical step to extend this to the acceptance of a nondeterministic component in the human psyche which is responsible for the ultimate decisions which govern the actual behavior of man in this world subject only to the boundary conditions of probabilities. Thus, in the spirit of quantum mechanics we can predict that most of the time the secularist will be oblivious to the yearnings of his inner spiritual essence, but we must also admit the possibility of the leap of faith, the complete change of heart and direction in life that can make him once again respond sensitively to his Divine soul. It seems that all the sophisticated partial differential equations of modern physics will never be able to ultimately explain the richness of the human spirit, the infinite spiritual capacities of man, and the potential for a true return to the inner life of the spirit.

The elements of hubris in the classical mentality are perhaps best exemplified by a famous anecdote.[7] Napoleon Bonaparte, upon receiving a copy of the *Mechanique Celeste*, the astronomical magnum opus of Laplace, is reported to have remarked: "You have written this huge book on the system of the world without once mentioning the Author of the universe." To which the eminent mathematician replied: "Sire, I have no need of that hypothesis." With the passage of the centuries, as man has refined his knowledge of the universe around him and simultaneously rediscovered the limits of his understanding and power, he has once again come to the recognition of the need for "that hypothesis."

## MATTER–ENERGY DUALITY AND THE
## UNDERLYING UNITY OF PHYSICAL REALITY

In classical mechanics, a basic distinction was maintained between matter and energy. The various manifestations of energy (electrical, chemical, thermal, gravitational, etc.) may be transformed into one another as may the various states of matter. However, the realms of matter and energy remain entirely disjointed; each realm retains its own integrity and is subject to its own conservation law. This kind of duality of the physical (matter) as separate from the analogue of the spiritual (energy) is central to classical physics. In contrast, a cardinal tenet of modern physics is the complete unity of the universe. Matter and energy are just different manifestations of the same underlying physical reality. As predicted by Einstein's special theory of relativity, matter can be transformed into energy and vice versa. This synthesis is even more dramatically demonstrated by the theory of quantum electrodynamics. The picture which quantum electrodynamics portrays of the underlying ground of reality—of the very nature of space itself—is profoundly different from the static conception of classical theory. The universe is seen as continuously involved in transformations in which matter and energy are spontaneously created and destroyed. Matter in the form of particle–antiparticle pairs instantaneously comes into being and disappears. Although this fundamental dialectic is not directly observable, its manifestations become apparent in such phenomena[8] as vacuum polarization, Zitterbewegung, and the Lamb Shift—phenomena which make quantum electrodynamics one of the most precisely verified theories in all of physics from an experimental standpoint.

## RATZO V'SHOV

This picture strikingly parallels the Torah perspective. In Jewish mysticism the ultimate dialectic of the physical universe is described as a continuous process of *Ratzo* and *Shov*.[9] *Ratzo* signifies the mystical union of the finite with God—the loss of identity, sense of self, and ultimately voiding of the physical which accompanies a spontaneous plunge into the Infinite Transcendent Source of the universe. *Shov* signifies the return to physical reality, the coming back down into the material universe as a differentiated entity brought into being by the dictates of the Divine Will. Although the Torah perspective provides a far deeper insight into the dynamics of the process described (Divine purpose as opposed to the spontaneous inexplicable fluctuations in the void of quantum electrodynamics), at the phenomenological level the descriptions of this ultimate dialectic are remarkably close.

Of course, the ultimate unity of all things as manifestations of Divine Will is the central image of Jewish mysticism. Just as the potential for the infinite, the abstract, the amorphous—the spiritual and its physical analogue—energy emanates from Him, so also does the potential for the finite, the limited, the categorized—the material. Thus, the physical and spiritual—matter and energy—are both manifestations of the Divine Will underlying reality and hence can be freely interchanged and transformed. Here, also, we have an almost exact parallel to the matter–energy duality of special relativity.

## SUBJECT–OBJECT DUALITY

A further instance of duality which pervaded classical physics is the sharp differentiation made between the observing subject and the observed object. To classical mechanics, man—the subjective observer—can be idealized as being wholly apart from the object of his observation. His interactions with this object are incidental to the observing process. Man's internal, subjective life is disparate from the external, objective reality of the universe around him. This is in direct opposition to the quantum mechanical view.[10] In this view such a duality no longer obtains. The observer (subject) and observed (object) can be described only as parts of a total all-encompassing system. The process of observation itself alters the state of the system—the conditions of the very thing to be observed. Internal life and the external universe—man and his environment—constitute one indissoluble entity. Any idealized separation, any duality, so distorts the actual situation as to make the resulting system meaningless.

In Jewish mystical philosophy there is a similar profoundly holistic image of man as part of the plenum of reality. To some extent this is expressed by the microcosm–macrocosm apposition in which the universe—the macrocosm—is regarded as a reflection and manifestation of the archetype—man—while simultaneously man reflects and manifests the structure of the universe.[11] The theory underlying this reciprocal relationship is that every aspect of the universe is a revelation of Divine creative energies. Hence at every level of the cosmogenic process there is expressed the pattern of the same primal Divine creative energies—a "homeomorphism"—at the levels of the individual, human society, and the totality of the universe.

## OUTER AND INNER ENVIRONMENT: A REFLEXIVE RELATIONSHIP

Another major way in which Jewish mysticism vitiates the artificial subject–object distinctions is in its representation of man as the internalizer of his environment.[12] Man cannot remain a separate, objective observer. He internalizes all external experiences to which he is exposed and incorporates them into his very being. When exposed to the potential for evil in the world man does not remain aloof. He assimilates part of that evil within himself, thus making his efforts to overcome his grosser nature that much more difficult. The nature of the environment within which man lives profoundly affects his perception and understanding of the world around him. On the other hand, man's actions—the manner in which he comports himself in his world—profoundly affect, on a physical as well as metaphysical level, the very nature of the world around him.

In marked contrast to most mystical systems, Jewish mysticism is profoundly action oriented. The most exalted flights of metaphysical speculation, the most sublime states of ecstatic mystical union, are a valueless perversion of man's purpose if they are not coupled with an *uvkhein*—a constructive, practical consequence with respect to his life and relationships in the here and now, the material world around him. Whether it is to sensitize him in his relationships with his fellow man or to reinspire him to higher devotion to his Creator through his actions (*mitzvot*) in this world, some positive behavioral modification is of crucial importance. Man, the internal subject, and his universe, the external object, must become a synthesized whole.

## CONCLUSION

I hope this modest effort has at least begun to demonstrate how, after centuries of secularism, modern science is once again beginning to come back to its ultimate roots—to a perception of reality greatly in concord with traditional Jewish beliefs and ultimately perhaps to some appreciation of the workings of the Grand Author of all creation. One question remains: What *uvkhein*—what practical consequence— can we, as Jews, derive from this realization? How can we utilize these insights to inspire and perfect ourselves and the world around us? How can we utilize this Divine manifestation through the natural order to bring ourselves closer to God through prayer, learning Torah, and performance of the Commandments? This question must be answered by each of us.

From *B'Or Ha'Torah* 1, 1982.

# Biology

# 10
# Living Water: Concept of *Midah ke-neged Midah* and Cellular Homeostasis

## Paul Goldstein

God has endowed all things with a divine spirit that originated at the time of Creation and includes even individual cells that make up our body along with regulatory mechanisms that keep different types of cells working in concert with each other. This interaction is known as the phenomenon of homeostasis, which is the very basis of life.

Homeostasis means that the internal environment of all living creatures remains relatively constant regardless of the conditions in the external environment. Thus, the activities of all tissues and organs must be regulated in a coordinated fashion such that any change in the internal environment (of the cell) initiates a reaction to minimize the change. Homeostasis works via compensating regulatory responses (measure for measure), and such responses require continuous sampling and correction.

There are essentially two types of regulators: intracellular (within the cell) and extracellular (where the influence is derived from outside the cell). Each individual cell (which can be equated with a person) exhibits some degree of self-regulation (intracellular), yet the

135

existence of many diverse cells forming numerous organs within the body obviously imposes the need for an overall regulatory mechanism to coordinate the activities of all cells. To accomplish this, extracellular regulators must establish communication between the diverse body parts. This is realized by, among other factors, the nerves that emanate from the brain and blood-borne chemical messengers called hormones (see below).

The *Midrash* often discusses the idea of *midah ke-neged midah* (measure for measure in equal terms) whereby a person receives reward or punishment from God based upon his own deeds. Some of the notable examples that illustrate this point are:

1. Titus entered a place where he did not belong, i.e., the Holy of Holies. Consequently, God caused a gnat to enter his brain, a place where it did not belong.
2. Abraham's descendants were rewarded by God in the wilderness with manna, the Well of Miriam, quails, and Clouds of Glory because of Abraham's hospitality to the three angels.
3. Serach revived Jacob's soul by telling him that Joseph was alive and well in Egypt, thereby causing *ruah ha-kodesh* (the holy spirit) to return to him. For this, she was rewarded a special place in the world to come.
4. Joseph received ten years' imprisonment in Egypt for speaking against his ten brothers. Yet, he was granted rulership by God for his refusal to touch Potifar's wife. Since he refused to listen to her, all of Egypt later listened to his command.

Are these apparently different concepts, *midah ke-neged midah* and homeostasis, related to each other? If we examine some of the basic forces of our life we can see how the idea of *midah ke-neged midah* operates at the cellular and regulatory levels.

## OSMOSIS

Osmosis involves the flow of water through the cell membranes of all living organisms to equalize the concentration of ions on both sides of the membrane. If a negative ion flows in, a positive ion flows out (if equilibrium is desired). Thus there is replacement of ions, measure for measure, to maintain a healthy cell. Humans, for example, are osmoregulators, which means we maintain an internal osmolarity different from our external environment. Thus, even at the cellular level we safeguard our existence. Cells cannot tolerate drastic changes in extracellular osmolarity and depend entirely on the osmotic regulation of the extracellular fluid. This extracellular fluid is *mayim hayyim* (living water), and the Torah is living water. Thus as Torah helps keep our lives in order, *mayim hayyim*, via osmosis, maintains cellular function.

## HORMONAL REGULATION OF GLUCOSE LEVELS

Insulin and glucagon are hormones produced in the pancreas that regulate the permitted level of glucose in our blood. Insulin is secreted when blood glucose levels are elevated and stimulates increased uptake of glucose by liver, muscle, adipose tissue, and other cells, which results in a rapid decrease in blood glucose levels. When fasting, blood glucose levels would fall too low, causing injury to the body, if not for the action of two hormones which are produced in response to decreasing blood glucose: glucagon from the pancreas and epinephrine from the adrenal gland. Breakdown of stored fat by the combined action of glucagon and epinephrine, complexed with cyclic AMP, results in providing the brain with the glucose that it requires to continue functioning. Through the actions of these two hormones the concentration of glucose is maintained within a narrow limit (1%) regardless of whether the person has just eaten or has fasted for several days. The hypothalamus in the brain also regulates blood glucose levels, particularly in times of stress, and produces a response called "hunger," which makes us want to eat and restore energy resources. If any of these regulatory mechanisms are not heeded, then diabetes may result and the rate of facilitated transport of glucose is decreased.

Of course, there are times in the life of a cell in which equilibrium is not desired, for example, during cellular secretion. In addition, nerves operate by receiving impulses, or action potentials, that rise and fall. However, this becomes part of the natural flux of the cell cycle, and its activity is controlled within its own limits. Ultimately, numerous cellular functions are controlled by the brain,[1] which contains specific centers that regulate factors such as nerve response, temperature, and blood pressure. The brain is a control center which in turn is regulated by the soul. When our souls are regulated by Torah—through our *hokhmah* (wisdom), *binah* (understanding), and *daat* (knowledge)—then we live in the light of the Torah.

From *B'Or Ha'Torah* 5E, 1986.

# Medicine
# and Public Health

# 11

# Organ Transplantation and Jewish Law

## Abraham S. Abraham

Organ transplantation raises a number of complex problems in *halakhah* (Jewish Law), and here I shall attempt to briefly discuss the various issues as they affect both potential donor and recipient. This is not meant to be definitive or authoritatively binding, but rather presents a discussion of some of the major problems with reference to various halakhic authorities who have expressed their opinions on the issues raised. Four major aspects will be discussed: (1) the recipient; (2) the live healthy donor; (3) the cadaver donor; and (4) the dying donor.[1]

## THE RECIPIENT

Whether a particular form of transplant is halakhically permissible or not depends on the mortality risk of the procedure and on the chances of its prolonging the life of the patient over and above the expected life span without the transplant. These considerations obvi-

141

ously include not only the results of world medical experience with the particular form of transplant under consideration but also the experiences and results of the particular center performing the transplant.

## THE LIVING DONOR

The basic halakhic question here is whether one may place oneself in a situation of possible danger to life in order to save another person who is in certain danger of life. The problem is discussed in the Jerusalem Talmud.[2] Although there is an opinion that one is obligated to risk one's life in order to save that of another, the *Shulhan Arukh* (the Code of Jewish Law) is silent on the issue. This has led later authorities[3] to deduce that the *Shulhan Arukh* does not accept the view that there is an obligation to risk one's life in order to save another. A sixteenth-century sage, the Radbaz,[4] concludes that where the danger to life of the "donor" is great, one is not obligated by law to take this risk. However, if the risk to life is small, then one is obliged to take this risk in order to save one's fellow man. Later, authorities have stated that if the risk to life is more than minimal, one is not obligated. However, one may, if one so wishes, take the small risk involved;[5] and our sages warn us not to be egoistic and selfish in our reactions to such a situation.[6]

1. *Renal Transplant.* Renal transplantation has proven to be a safe procedure from the point of view of the donor. Thus, of 7,000 donors, there was an operative mortality of 3.[7] Nor is there evidence of any serious problems with renal failure over long-term follow-up.[8] Therefore, on medical evidence most contemporary authorities[9] agree that it is permissible and even praiseworthy for a healthy relative to donate a kidney in order to save the life of a patient with terminal renal failure. However, Rabbi Weiss[10] remains doubtful about the basic question of whether one may risk one's life to save that of another.

2. *Bone Marrow Transplant.* In bone marrow transplant the only risk to the donor is that of general anesthesia, and probably all authorities would agree that such a transplant would not only be permitted but indeed be praiseworthy.

## THE CADAVER DONOR

The halakhic problems of removing the cornea after death revolve around the questions of (1) deriving benefit from the dead; (2) desecration of the dead; and (3) nonburial.[11]

1. Most authorities agree that it is prohibited by the Torah to derive any benefit from the dead or from any part of a corpse unless it is necessary to save a life. However, Rabbi Unterman[12] has suggested the ingenious idea that since after transplantation the cornea is "brought back to life," one may now view it as a living organ; the benefit to the recipient is therefore from a living and not a dead organ. Rabbi Tzvi Pesah Frank[13] also permits the reception of a corneal transplant as does Rabbi Moshe Feinstein.[14]

2. According to Rabbi Ovadiah Yosef,[15] though desecration of the dead is obviously prohibited, "desecration" means purposeless mutilation, namely, where there is no danger to the life of the recipient. Therefore, restoration of sight (which is tantamount to saving life) to a blind person by corneal transplant would be permissible (even for blindness in only one eye).

3. Rabbi Yosef[16] states that according to many authorities the obligation to bury the body as a whole organism probably does not apply to such small portions of body tissue. However, other authorities[17] disagree. They argue that while the dead are under no moral or ethical obligations toward the living, the duty of burying the dead is overriding and includes every part of the body. Rabbi Waldenberg[18] concurs but adds that if as a fait accompli the cornea has been removed and is available, it may be transplanted.

According to Rabbi Shlomo Zalman Auerbach,[19] cadaver skin transplant in order to save the life of a patient suffering from burns is permissible.

## THE DYING DONOR

The dying patient, no matter how close to death, is halakhically alive.[20] The problem of transplants is obviously dependent on the definition of "death" and the differentiation of the dying from the dead.

The earliest source is the Talmud,[21] where cessation of respiration is held to be of decisive importance. Thus, in the absence of respiration, no examination of the heart is needed in an unconscious individual in order to establish the state of death.[22] However, one must remember that a patient with respiratory arrest in those days could probably not be resuscitated, and knowledge as to whether the heart was still beating or not was therefore immaterial. But Rabbi Moshe Sofer[23] decided that only an unconscious individual who has no respiration and no heartbeat may be considered dead. Rabbi S. M. Schwadron[24] also points out that cessation of respiration is by itself not sufficient to pronounce an individual dead if there are signs of life in any other part of the body. Present-day authorities[25] lay down the same conditions.

Since then, however, the concept of brain death, and in particular the concept of brain-stem death, has come to the fore in medical practice, and in many countries brain-stem death (defined under strict criteria) is accepted as defining death of the comatose patient on a respirator, even in the presence of a beating heart.[26] Present-day halakhic authorities are divided as to whether brain-stem death is an acceptable halakhic definition of death. Although the Chief Rabbinate of Israel has, under very strict and limited conditions and criteria, permitted such a patient to be used as a heart donor,[27] Rabbis Eliashev, Vozner, Waldenberg, and Weiss refuse to accept this definition.[28]

From *B'Or Ha'Torah* 7E, 1991.

# 12

# Ethical Issues
in Community Health

## Velvl Greene

### INTRODUCTION: AN AWAKENED
### INTEREST IN MEDICAL ETHICS

Codes of medical ethics and standards of behavior for the medical practitioner are part of the traditional fabric of medicine. It is with some surprise, therefore, that we note the recently awakened interest in medical ethics. Books on health ethics are flooding the market. Seminars, symposia, institutes, and workshops devoted to the subject are organized whenever financial sponsorship can be found. Very few medical schools or schools of public health are without their resident expert in ethics.

This awakened interest results in part from the sensitivity of conscientious physicians to the many nonmedical issues that influence medical practice, such as law and economics. Dissatisfaction has also grown with what passed for medical ethics in the past, which confined itself to topics such as commercial advertising by physicians, sexual improprieties between practitioner and patient, and the misuse of prescription narcotics. Without doubt, the main reason for the renewed interest in medical ethics is the unprecedented advance in medical technology of the last few decades. The role of the doctor has expanded from healer to one who can decide whether to prolong or terminate life. The power to intervene in human life has become routine through new technologies such as organ transplant, in vitro fertilization, genetic engineering, and electronic life support. This technology has done for physicians what nuclear weapons have done for governments—given them a power which outstrips their wisdom to use it.

## RESOLVING MEDICAL ETHICS
## IN A SECULAR WORLD

The physician is trained to apply technology correctly. In this context, "right" and "wrong" refer to whether a procedure or drug produces the expected outcome. The larger questions, of whether the technology should be used at all, or who should be treated first when a conflict of needs arises, have been reserved for the ethicist.

The ethicists aren't doing very well, to my eye. Although their conferences, books, and symposia may appear in the newspapers, their impact on the practice of medicine is negligible. They have articulated important questions but they haven't come up with answers. In this respect, health ethics doesn't differ from other fields of secular ethics, such as business, politics, or war, in which little progress has been made. The major criterion of right and wrong still seems to be whose financial interest is at stake.

Of course, medicine is practiced according to rules and standards. However, these have been created not by ethicists but in the law courts and legislatures. The practice of medicine often calls for quick decisions which cannot wait for the resolution of an esoteric debate which may have gone on for centuries. If the decision is ultimately going to reflect community consensus, then the attorneys, farmers, and businessmen who sit in our legislatures should be considered as ethical as doctors or professors.

Whether these decisions are motivated by the basic aim of ethics— to do the "right" thing in a situation in which choices are presented— is not immediately relevant. Laws have been passed, administrative edicts formulated, and court decisions rendered that effectively regulate the practice of medicine. Legal guidelines define the beginning and the end of life, as well as the responsibilities of doctors and hospitals in prolonging or inhibiting life. Ample legal precedent exists to permit state agencies to preempt the role of parents if a state attorney can convince a judge that the lives and health of minor children are at risk. The law defines, with remarkable precision, the conditions and even the environment in which pregnancy can be terminated and by whom. The law defines the time a mass of cells becomes a person and defines this person's rights to treatment at different stages of development thereafter. Terms such as professional competence, informed consent, negligence, and life support measures, which defy any quantitative definition, are commonly defined in courtrooms. In other words, medical questions which have been the exclusive property of ethics and theology for generations have been resolved in operational terms by legislative and judicial procedures.

Many of these issues have become highly politicized. Political action groups are formed to support or oppose such issues as abortion, sex education, and the supplying of contraceptives to unmarried people. Victory is less a matter of ethical principles than of organization, fund raising, and lobbying skill. Abortions were once illegal, presumably because they were unethical. Doctors who performed abortions were arrested, disgraced, and suspended from practice. Today, abortions are not only legal, but they are promoted. Presumably it is unethical to keep an unwilling woman pregnant. But if the right-to-life lobby can influence the composition of the next Congress, abortions will become illegal, and once again an abortion will be considered more unethical than keeping a woman pregnant against her will.

Other medical issues gathering lobbies are the right to euthanasia and the merit of non-orthodox healing practices once called quackery. Quackery used to be considered unethical and was against the law. But homosexuality was also illegal once. Today the Gay Caucus of

Public Health Workers is an officially recognized body within the American Public Health Association. This Gay Caucus has annual meetings, sponsors educational programs, and lobbies for more funds for AIDS research. Obviously, medical ethics are not permanent and immutable.

The recent "Baby Doe" controversy is an excellent case history of an ethical problem resolved by the judiciary. The case pitted the Department of Health and Human Services against a New York hospital, its doctors, and the parents of a baby born with severe congenital defects. Was the hospital responsible to maintain an unwanted little girl on life support who had only a slim chance of surviving, a prospect of a dependent and miserable life if she did? Characteristically, this case was tried in the courts. Although the experts in ethics are still debating the issue, the doctors couldn't wait for their decision. They were compelled to obey the legal decision—to keep the baby alive by artifical means.

## JEWISH MEDICAL ETHICS

The Jewish world is also experiencing a new interest in medical ethics. In the thirty years since Sir Immanuel Jakobovits published his magnum opus on Jewish medical ethics, we have seen a literal explosion of books, journals, and monographs on the subject. Of course, the ethical challenge posed by the new technology affects the Jewish medical system as well. Although the questions are the same, the difference lies in the answers and their source.

Actually, Jewish concern with medical ethics stretches to the early beginnings of Jewish history. A review of Rabbi Jakobovits's volume will impress anyone whose knowledge of medical ethics is confined to modern non-Jewish publications. Although the narrative reads like a debate among contemporary scholars, a glance at the appendix shows the reader that the debaters span several thousand years. No opinion is considered outdated or too modern.

Even more exciting is the recent translation from the German of Professor Preuss's classic on biblical and talmudic medicine published originally at the turn of the century. There the foundation of Jewish medical ethics is clearly and systematically outlined. Since the observant Jew believes, with perfect faith, that the Torah, both written and oral, is the revealed instruction provided by God to guide every aspect of human life, the references in the Torah to medicine and health represent the eternal values that must guide behavior in those areas. There can be no dichotomy between ethics and Torah because they are identical.

In the Jewish context, ethical decisions are not made by juries or legislatures. Instead, the Jew depends on a responsum from a qualified rabbi, one expert in Talmud, the legal codes, and the responsa that have been handed down over the generations. The Talmud provides the philosophical and theological foundation, the codes provide the legal systematics, and the responsa are the precedents. Based on these sources, a God-fearing *posek* (adjudicator) will derive a practical answer acceptable in Jewish law as a continuation of the *halakhah* (law) that originated with the Almighty's instructions to Moses. This responsum cannot be taken lightly. Halakhic decisions rendered by the unqualified are the equivalent of brain surgery performed by the untrained. In Jewish tradition, the responsa become a part of Torah and represent divine instruction.

In Jewish life there have always been halakhic controversies which were resolved by the most eminent *posek*. Ultimately, one body of law evolved. Because Torah is the source of our ethics, the physician who practices according to halakhic decision is behaving ethically, by definition.

In the Jewish world, we don't search for an "ethical value system" on which to base a medical decision. When faced with an ethical problem, a physician must find a qualified *posek* and pose the question: "This is the medical situation, these are my alternatives, what does the Torah say?" The answer the *posek* provides will have been judged within the same framework as the rest of Jewish life, such as kosher food laws, business dealings, marriage, divorce, and *Shabbat*. And the answer will be an ethical one.

After centuries of responsa, certain principles can be derived.

These can be classified, as Jakobovits has done for medical ethics. However, the principles elucidated cannot be used to make halakhic decisions for oneself. Significantly, the publishers of most journals in *halakhah* include the following disclaimer: "To avoid any misunderstanding we would like to emphasize that no halakhic conclusions should be drawn from the material published here. Any actual halakhic problem should be brought to the attention of a qualified rabbi." Evidently, medical ethics are not to be taken or rejected at the whim of the consumer.

## MEDICAL ETHICS ISSUES COVERED IN CURRENT HALAKHIC LITERATURE

I am far from an expert in halakhah. However, even a layman with access to a library of medical-halakhic literature will be impressed with the wealth of subjects for which responsa are available. Once I prepared an index of these subjects for some medical students at Ben-Gurion University. This is what I found:

1. Legality of human acts of healing in a world where life and death are controlled by God
2. Recourse to prayer, irrational cures, and nontraditional medicine
3. Moral responsibility of the physician with respect to
   a. saving life
   b. arbitrary termination of life
   c. professional secrets and confidences
   d. patients' consent and access to information
   e. treatment of pain
   f. treatment under conditions harmful to doctor
   g. definition of disease, treatment, and death
4. Potential conflicts of *halakhah* and medical imperatives with respect to *Shabbat*, holidays, festivals, fasts, dietary laws, autopsy, prayer and sanctity in the hospital environment, and treatment of one's own parents

5. Reproductive problems such as fertility, eugenics, sterilization, contraception, abortion, artificial insemination, and in vitro fertilization
6. Surgery, transplants, and organ donation
7. Circumcision and *pidyon ha-ben* (redemption of the first born)
8. Laws of mourning
9. Visiting the sick
10. Care of the body and limbs after death
11. Psychiatric and social medicine
12. Child abuse.

This list is far from exhaustive. Any scholarly review of the literature would reveal many other topics with ethical implications.

One major area of health, however, is conspicuously absent: the ethical issues related to public health. Almost all of the responsa I have encountered, and most of the review articles and books, are devoted to the one-on-one interaction between physician and patient. Nearly every answer is devoted to the problems of the sick, the dying, and the dead. Few responsa deal with public policy, community health programs, and government interventions.

This is understandable, considering the mechanism by which responsa are generated. The *posek* is answering a specific question posed by an individual—a doctor, patient, patient's family, or hospital functionary. Public health officials and government bureaucrats hardly ever consult *poskim*. Moreover, the ones who ask the questions are usually concerned with an illness or death after it has occurred.

The Talmud and the codes are full of information about health preservation and disease prevention. Again, these instructions are intended for individuals. There is not much available in halakhic sources to guide those responsible for community health.

I submit that there are critical ethical questions in public and community health which need the attention of halakhic scholars. Since the Jewish community is already engaged in community health activities, the disregard of *halakhah* is itself unethical. Perhaps this essay will be the initial step in a much-needed dialogue between the public health professional and the *posek*. From this dialogue could emerge proper guidelines based on Torah for the resolution of public health issues in much the same way as we have for traditional medical practice.

If not, we will revert to the mechanisms used today—legislation, litigation, and administrative fiat, supplemented by the occasional statement from an ad hoc committee whose members are oblivious to or ignorant about the ethics of Torah.

## AN INTRODUCTION TO COMMUNITY AND PUBLIC HEALTH

Most people, many health professionals among them, will probably be surprised at my distinction between medicine and public health. The prevailing impression is that physicians are the essential health providers, that hospitals are the essential health institutions, and that public health is simply the end result of their effectiveness. Actually, doctors, nurses, and pharmacists constitute only a fragment of the larger picture of community health. Health is much more complex than the absence of disease. In fact, The World Health Organization defines health as nothing less than "the complete physical, mental and social well-being of a person." It follows that the term *health provider* identifies not only those physicians and nurses who care for and heal the sick but also the educators, engineers, sanitarians, administrators, psychologists, nutritionists, and many others.

While classical medicine deals mainly with the sick, community health deals with the total population, the sick as well as the healthy. Classical medicine is concerned with the cure of disease, while community health is concerned with the prevention of disease before it occurs, as well as with the individual who cannot be cured.

On the practical level, medicine and community health are interwoven. Physicians are as interested in preventing disease as they are in healing. But the scope of community heath goes considerably beyond the simple triad of doctor, patient, and hospital. If vaccines are necessary to prevent polio, we don't wait for patients to visit their doctors. Instead, we provide the vaccine to school nurses, make it available in neighborhood fire departments, and provide vaccination booths at state fairs. If traffic accidents are the leading cause of death and disability among children, we design better highways, legislate compulsory seat belt use, and arrest drunken drivers. If we recognize that the physical, mental, and social well-being of the poor is impaired by malnutrition, we try to remedy the underlying problem by giving charity, establishing school lunch programs, or distributing food stamps. All of these activities are within the scope of community health.

One branch of community health is called public health. It refers to a variety of health programs and services administered by governments. Historically, public health agencies supervised such enterprises as infectious disease quarantine, water supply safety, sanitary sewage disposal, and solid waste management. Later, programs included health education, vital statistics registration, visiting nurse services, environmental pollution, and health care financing. The actual scope varies from state to state, locality to locality, and country to country. The unifying characteristics of all public health agencies are threefold:

1. They rely on the government's tax revenues to finance the health program. Every taxpayer contributes, whether he needs or wants the program or not.
2. They rely on the police power of the state for enforcement and compliance. Although every purveyor of food and drugs should naturally strive for their purity and safety, those who don't will be punished by the loss of a license, a fine, or jail.
3. They involve bureaucratic processes, with all of the attendant benefits and limitations.

The history of public health in the last one hundred years makes fascinating reading. The health revolution in America and Europe of this century can be measured by a historically unprecedented drop in infant mortality, increase in life expectancy, and virtual conquest of infectious disease. Scholars attribute this revolution almost entirely to the direct intervention of public health agencies in the lives of nearly everyone.

The role of public health agencies is still expanding, in administrative, legislative, and judicial processes that have a direct bearing on health. In addition, governments exert their impact on health through other agencies that deal with agriculture, ecology, welfare, education, commerce, and finance.

Clearly, governmental decisions have as much, if not more, influence on our health as do decisions made by doctors and hospitals. If ethical standards of behavior are being established for medical practitioners, similar standards should be established for the bureaucrats whose actions influence public health.

From the Jewish point of view, what does the *halakhah* have to say about the dilemmas we face in public health? As an agenda for discussion, I have selected those topics for which the need for guidance seems the most critical:

1. Public policy and public health priorities
2. Health care costs and economics
3. Quality control of health services
4. Human experimentation
5. The government as regulator
6. Individual rights versus community responsibilities.

## PUBLIC POLICY AND PUBLIC HEALTH PRIORITIES

In the usual one-on-one relationship between doctor and patient, the subject to which most medical *halakhah* is devoted, the doctor is only rarely forced to neglect one patient in order to treat another. At any given moment, the doctor's full attention and resources are devoted to the patient in front of him.

In the field of public health, the patient is replaced by a population, an aggregate of individuals, each with unique problems, needs, and expectations. By their very nature, public health programs must address some problems while neglecting others. For example, an infectious disease control program will benefit mainly those people who are susceptible to the disease. Similarly, a breast cancer surveillance program is a matter of life and death to women of a certain age and background. It is of limited value to the health of males in the same community. Yet everyone pays for the program, even those who will not benefit.

As a consequence, a key ethical problem in public health is the establishment of priorities. What health problem is the most important? Whose problems do we attack first? How can limited community resources be allocated equitably? How do we use the considerable powers of the state for the greatest community health benefits?

Of course, sometimes a doctor must also make priority decisions. In an emergency room, it might be impossible for one physician to pay equal attention to all who need his help. If he treats patient X first, patient Y might die. This classic scenario has a name: triage. In public health, triage is a daily problem.

Triage decisions in medicine must also be made when rare therapies or a limited number of medical devices must be allocated among many patients. This difficulty arose a few years ago, when hemodialyzers were rare and endstate renal disease patients were numerous. It occurs today in transplant surgeries when several patients are waiting for the next kidney suitable for transplant. Who should get it? The philanthropist who has endowed the hospital, or the young mother who can't afford the operation? These ethical questions are being debated in today's secular ethics and in current halakhic responsa.

Less dramatic scenarios, but with equivalent ethical consequences, are the basis of today's public health debates. How many hospitals will we need in the next twenty years? What shall we do first, build a water purification or a sewage treatment plant? Where shall *Kupat Holim* (the Israel sick fund) establish the next public clinic? Should we enforce the air pollution laws even if the factory which employs hundreds of workers will be forced to close? Shall we vaccinate all of our children against measles, at great expense, or hope that there will be no epidemic this year? Can we force doctors to report all cases of syphilis and thereby interfere with the sacred doctor–patient trust? Can we forcibly hospitalize all cases of tuberculosis?

Priorities in public health go far beyond the provision of direct and indirect medical services. For example, how should a nation allocate medical research funds? If we spend a billion dollars in the fight against cancer, some of that money, research talent, and laboratory facilities could have been used to fight other diseases. Is cancer more important than heart disease, schizophrenia, or spotted fever? In a democracy, decisions like these are made almost weekly by legislators and bureaucrats. Are their decisions halakhically valid?

In matters of immediate life and death, priority decisions are relatively easy to justify. Mistakes can sometimes be forgiven because of the urgency of the moment. In public health, there is usually time to calculate the costs, risks, and benefits, but the decisions are no easier to make. Although the consequences of the decision may be delayed, the cost must still be paid. For example, the introduction of malaria control measures in Southeast Asia saved literally millions of children from inevitable premature death. Now these children are adults with children and grandchildren of their own. The population of that part of the world is rapidly outstripping its food resources. Is death for some from starvation preferable to death for others from malaria? What if a solution for low birth weight is discovered, one that will increase the population even further? How would the *halakhah* address this problem?

In another case closer to home, Israeli public policy decisions regarding sanitation and medical services in Gaza since 1967 have dramatically curtailed infant mortality rates among the Arab inhabitants. From a public health point of view, any program that lowered infant death rates from 120 per 1,000 live births to 40 per 1,000 must be considered a great success. However, Arab population growth presents a grave political problem for Israel. Does the *halakhah* permit us to use or withhold health programs to achieve political goals? Would it be ethical for Israel to provide anything less than the best medical service to any population within her jurisdiction? Is it ethical to provide clean water and sanitary sewage disposal for some and not for others?

# HEALTH CARE COSTS AND ECONOMICS

Many volumes of Talmud and innumerable responsa deal with the *halakhah* of financial transactions. These decisions provide guidelines for the fees paid privately to doctors and hospitals. In most civilized countries, however, the system has changed from direct private payment to indirect public payment. In Israel, more than 90% of the population is enrolled in the various sick funds like *Kupat Holim*. England has operated a system of socialized medicine for forty years. Even in the United States, the bastion of private enterprise, more than two-thirds of the three hundred billion dollars spent last year for health care was paid for by third-party agencies: the government, health insurance underwriters, and various prepaid medical plans. The shift to indirect payment brings new ethical questions in its wake, simply because any individual or group which collects and dispenses someone else's money is confronted with decisions about fairness, priorities, and accountability.

For example, when England inaugurated its National Health Service after World War II, everyone was obligated to join. Since the service was funded by taxes, payments into the system were determined by income level. However, provision of medical service was predicated on need. This led to the inequity of some paying for less than they received while others were receiving more than they paid for. This is good socialism, but is it ethical? Conversely, is it ethical for the rich to receive better medical treatment than the poor just because they can afford it? The question can be posed in a different way: Is good health a privilege or a right? How does one define good health care? Is there a threshold between good, adequate, and less than adequate health care? Is this an ethical or a medical decision?

It has been suggested that Jewish answers to such questions might be derived from the *halakhah* that regulates charity, a subject thoroughly treated in Jewish sources. However, government health care financing agencies are not charities. The donor is not asked to contribute but is compelled to do so by law. He is certainly not consulted about the amount that must be paid. The sum is deducted from his salary before he even sees it and he has little to say about how the funds should be spent. Very few legislators or bureaucrats consult *poskim* about the ethics of collecting or dispensing such funds.

Between the private and governmental payment systems, there exists the middle world of health insurance underwriters and prepaid health plans like the *Kupat Holim* and the Health Maintenance Organizations. At least these systems permit freedom of choice. The consumer can decide whether or not to join and can shop among the alternative plans. But ethical problems arise even here. The plans and insurance options must operate in a fiscally responsible manner or they become bankrupt. Unfortunately, the words "fiscally responsible" are too often code words for restriction of services or exclusion of high-risk clients. Ultimately the management of third-party agencies must make the same ethical decisions as the government administrator: What level of health care can we afford to pay for? How do we deal with patients who abuse the system? Whose criteria should be used to evaluate adequacy of service—the doctor's, patient's, or administrator's?

There is yet another area of health-related financing that requires ethical guidelines: the activity of voluntary and charitable health organizations. On the surface, voluntary groups such as the Heart Association or the Arthritis Foundation are the epitome of human kindness. In reality, many of these organizations have become big businesses and bureaucracies in their own right. They collect millions of dollars, have efficient public relations operations, and lobby effectively for special favors from legislatures. Their focus is a given health problem rather than the overall health of the community. Since contributions to such groups are tax deductible, they actually compete with governmental agencies. A voluntary dollar contributed to cystic fibrosis research means fewer dollars for research in other fields that the government thinks are important. Moreover, these groups compete among themselves for the available charitable dollars. The Jerry Lewis telethon raises eight to nine million dollars each year for muscular dystrophy victims. The Sickle Cell Foundation doesn't have a charismatic comedian available and raises much less. Is that ethical?

## QUALITY CONTROL OF HEALTH SERVICES

The quality of health services will depend on the quality of the health practitioner. It is expected that the quality of a doctor will reflect the quality of the student admitted to the freshman medical class. Unfortunately, the selection of suitable medical school candidates is not an exact science. Balances must be struck between academic ability, personality, and promise. Striking such balances involves ethical as well as administrative input. Are we wasting a national resource by refusing admission to some who are qualified? After all, the ones who nearly made it are really as competent as the ones who just made it. In this respect, admission slots to medical school can be viewed as a rare medical resource, like dialyzers. How should they be allocated?

One would like to think that all physicians are equally competent and that all hospitals are of equal quality. All physicians with license to practice have undergone a rigorous training and have passed qualifying examinations, just as all hospitals are subject to inspection and licensing. Logic and experience, however, reveal that the quality of doctors and hospitals is variable. This variation can have an impact on life and death.

The community has the power to grant or to withhold licenses for medical practice. Usually doctors and hospital administrators set the standards for licensing, hopefully by objective standards. How does one balance the need for more doctors and lower prices against the equally compelling need to maintain the highest level of quality? If the opening of another hospital or the licensing of another dozen doctors means that the payment pie will be cut into more and smaller slices, can we trust those who will be harmed by the decision to be fair? Licensing and accreditation power has been a traditional weapon to restrict potential competitors; what is the ethical way to overcome this, when matters of health and life are involved? Another question of ethics arises in respect to the licensing board: How do we overcome racist and sexist bias, without discriminating against the competent members of the community who happen to be male and white?

## THE GOVERNMENT AS REGULATOR

Closely related to the matters of admission, licensing, and accreditation is the ever-growing role of governmental agencies as regulators in the health care field. At first glance, the ethical implications of this phenomenon might be unclear. After all, governments are morally and legally obligated to protect their citizens. Agencies that regulate hazards such as air and water pollution, pure foods and drugs, and toxic chemical and nuclear waste are really fulfilling the government's obligation. The restriction of certain private liberties for the safety and welfare of the community as a whole is within the power of any sovereign state. The ethical problem arises when the government starts to enforce such regulations arbitrarily, when regulatory decisions are based on irrational premises, and when subjective, aesthetic biases replace hard health data in risk–benefit equations.

In the United States, for example, a veritable alphabet soup of regulatory agencies has been spawned in the last twenty-five years to deal directly with matters of health. In addition to the well-known bureaus of the FDA and USDA, Americans have learned to comply with health-related edicts from the EPA, CPSC, OSHA, NRD, VA, ICC, CAB, FAA, and dozens of others. This is only on the federal level. States, counties, and cities all have their own health departments, pollution control agencies, and consumer protection bureaus.

No doubt the regulations promulgated and enforced by these agencies are well meaning. However, the sheer number of agencies and regulations has generated multiple conflicts between government and industry, between government and citizens' interest groups, and even between various government agencies. A new legal subspecialty has arisen: lawyers who specialize in governmental health regulations. There must be an ethical halakhic solution to these conflicts.

Three examples might suffice to illustrate such conflicts:

In the 1960s a congressman from New York amended the Pure Food and Drug Act in order to ban from food any additive that was carcinogenic to humans or animals. This is known as the Delaney Clause, or the Zero Tolerance Principle. It was a noble gesture twenty-five years ago and it still sounds noble today. However, in the last twenty years we have discovered that many natural foods, beyond the jurisdiction of this law, contain natural traces of carcinogens. This leads to the paradox of some foods in the marketplace being legal while others, which actually contain much less carcinogenic material, are illegal. In addition, our ability to detect carcinogens has improved tremendously. Today we have tests and instruments that are two to three orders of magnitude more sensitive than the ones available in the 1960s. This leads to the anomalous situation that food additives considered safe when the law was passed are now unsafe and illegal. The food is the same; only the technology has changed. Finally, there is the problem of interpreting carcinogenic assays. Mice fed high doses of saccharine for long periods of time develop tumors. There is no evidence, however, that saccharine is harmful to humans. The FDA banned saccharine in accordance with the law. Congress rescinded the ban in response to public outcry. What does the *halakhah* say?

The problems arising from governmental regulation can also be seen in the example of presterilized medical devices supplied to hospitals. Manufacturers label these products "for single use only" because they are afraid they will be sued if the hospitals reuse these items without proper sterilization. In addition, single-use items are quite profitable; after one use, an item must be discarded and a new one purchased. This can become a major problem for the hospital with a $200 intra-arterial catheter or a $5,000 pacemaker. The Federal Bureau of Medical Devices and the Joint Commission for the Accreditation of Hospitals insist that hospitals follow the manufacturers' instructions precisely. Many hospitals, however, have learned how to resterilize such items quite effectively. In many cases this has led to significant savings. The Health Care Financing Administration, which reimburses the hospital for Medicare and Medicaid patients, and the Veterans Administration, which operates several hundred hospitals, are quite interested in "re-use of disposables." The manufacturers' associations lobby against the practice as dangerous and unethical. What are the ethical restrictions? Can one ethically provide one patient with a brand-new device and another patient with a used one? Who pockets the savings—the patient, the hospital, or the third-party payer? What are the ethical constraints on spending public money unnecessarily, just because a manufacturer has labeled his product "single use"?

A similar problem arises from the goverment's effort to protect the public from quack cures for cancer. In our generation the prime example is Laetrile, a concoction made from ground apricot pits. After much testing and evaluation of clinical data, the FDA concluded that Laetrile is quite ineffective and banned it from interstate commerce. A large constituency of Laetrile advocates, including some doctors, claim that the FDA action was a ploy by the medical establishment to interfere with their liberty and health. They successfully convinced the legislatures of certain states to legalize Laetrile. What is the ethical position in this controversy? Should the federal law be enforced? What if it is found that patients treated with Laetrile could have been helped by more authentic therapies? What if no authorized therapy is effective? Can the law prevent the sick from grasping for whatever help is available?

## MEDICAL EXPERIMENTS
## WITH HUMAN SUBJECTS

The immediate ethical response to the words "human experimentation" is condemnation. In the context of this discussion, however, we will not address the experiments of the Nazis at Auschwitz or the "731 unit" of the Japanese Army. Instead, we are concerned with the ethics of modern medical research, supervised by compassionate and dedicated practitioners and conducted with the cooperation of the patients involved. The truth is that most useful medical discoveries involve some kind of field trial on human subjects. Polio vaccine was not licensed for general distribution until Salk completed his controlled studies of safety and effectiveness on several thousand vaccinated and unvaccinated children in the 1950s. The therapeutic properties of penicillin were proved by injecting human patients before it was employed to save the lives of thousands of wounded soldiers in World War II. Every surgical procedure we use today was once a "human experiment." The eradication of pellagra from the southern states had to wait until Goldberger learned to control the diets of inmates in Alabama prisons. Throughout the history of medical, surgical, and pharmacological progress, innovations have been tried in a clinical setting with human subjects.

There are ethical and less ethical ways to conduct such trials. Unfortunately, there are many cases of well-intentioned researchers doing quite the wrong thing while impelled by the noblest motives. How does one gain the acquiescence of the patient? What is suitable, informed consent, particularly if the patient is medically unsophisticated? Is it ethical to randomly sort patients into treatment and control groups, one of which will receive the treatment? Is it ethical, after the fact, to recriminate the doctor for having been put into the control group when the drug was shown to be beneficial? Is it ethical to allocate patients in any way but randomly? Is it ethical for well-meaning but untrained investigators to conduct clinical trials? How far into a trial should the investigator be "blinded" to prevent research bias if an indication arises that the treatment is either beneficial or harmful? In such a case, premature interruption of the experiment might obviate the trial and the significance of the results. Currently this area is under the intense scrutiny of secular ethics scholars and is a prominent topic of discussion at meetings of clinical researchers. It is, however, difficult to find it treated in any depth in the halakhic literature.

## INDIVIDUAL RIGHTS VERSUS COMMUNITY RESPONSIBILITIES

The halakhic literature dealing with the role of the individual in society is immense. However, when the interaction of individual and community is viewed from the narrower focus of health problems, the responsa become considerably more limited.

One example is the field of self-inflicted injury and illness. For example, in the case of venereal disease, except for the instances of rape, child molesting, and mental deficiency, nearly every person who contracts the disease is infected during a voluntary act. In the same way, although the victim of sunburn did not intend to become burned, he took his shirt off voluntarily. The glutton did not intend to become obese, nor the smoker to develop emphysema. In every one of these cases the voluntary act provided some kind of gratification. At the time, the risk was acceptable, considering the short term benefit. A gamble was taken, and the "victim" lost.

The victim then calls upon the community to provide treatment. The *halakhah* certainly sympathizes with the patient. But what does the *halakhah* say about the person who wants free treatment for venereal disease, sunburn, or obesity every time he or she gambles and loses? What control can the community exert to protect the victim against himself?

Rabbi Moshe Feinstein, the Dean of American *poskim,* has recently ruled that smoking is a life-endangering behavior and contrary to *halakhah.* For the first time in history, a number of *yeshivot* have acted to ban smoking in their study halls. This responsum will probably do more for the health and longevity of the young men affected than the billions to be spent next year on cancer research.

Another area of conflict between individual rights and community health responsibilities involves religious freedom. In America we encounter Jehovah's Witnesses who refuse blood transfusions, fundamentalist sects whose religious rituals involve the handling of poisonous snakes, and even some groups whose faith demands severe corporal punishment of their children. What should be the role of the state in these matters? Where does the state's concern with its citizens' health overlap with the right of those citizens to practice their religion?

I remember a case in one hospital where a patient dying from cancer was in extreme pain. The doctor compassionately prescribed generous doses of morphine to ease her last days. The patient was a devout Catholic and believed that the pain she suffered in this world was a form of expiation for her sins and that the more she suffered here, the brighter would be her heavenly reward. Accusing the doctor of interfering with her afterlife, she refused analgesics. Her screams disturbed the other patients. What should the Jewish doctor do in this conflict between individual and community?

Alcoholism is another difficult area. Once it was defined as a crime, then as a social problem. It is always a significant contributor to physical and mental illness. America tried to solve it once by prohibition, with little success. Can one take away the right of the entire population to have a drink just because a minority responds pathologically to alcohol?

How does *halakhah* deal with such diseases as AIDS? This is an often fatal consequence of a homosexual life-style, which is prohibited by *halakhah* in the first place. Are we permitted to spend public funds for treatment, research, or support counseling for AIDS victims? What does *halakhah* say about public funding of abortions in cases of rape and incest? How would *halakhah* rule on the ethics of using behavior modification to treat patients in state mental hospitals?

Some of these questions go to the very heart of public health philosophy and practice. Perhaps many of these questions have halakhic answers which I am not equipped to provide. On the other hand, I suspect that among these questions are some that have not yet been put to authoritative *poskim*.

I submit the preceding as a first draft of an agenda for serious consideration.

From *B'Or Ha'Torah* 5E, 1986.

# Evolution

# 13

# The Evolutionary Doctrine

## Lee M. Spetner

Only recently has it become possible to conduct a quantitative check of the probability of chance development, the basis of Darwin's suggested theory of random mutation and natural selection. The mathematical check carried out by the author shows that the probability of evolution by chance is many times too small to explain evolution. Furthermore, new biochemical data has shown that the amino acid sequences of the cytochrome-C protein of various organisms, hitherto considered to be identical, do not correspond. The many gaps remaining in the fossil records, along with the current consensus of opinion among biologists that microevolutionary change cannot cause macroevolution, further weaken the leading anti-Creation "theory" of modern science.

## THE "ETHICS" OF BLIND CHANCE

The evolutionary doctrine dominates the philosophy of the Western world. The assumptions and concepts of this doctrine pervade the thought and speech of the entire culture. Phrases like "primitive types" and "related organisms" are used by many people without an awareness that these terms make sense only when the evolutionary doctrine is assumed. No member of Western society is immune to these influences.

The doctrine of evolution says that all life stems from a single simple origin, and that life developed and branched in a purely random fashion from that origin during the course of billions of years to become all of the varied living forms that we find today on our planet.

There is no need to invoke a Creator who designed and created life; blind chance alone is sufficient as the prime mover of the development of life. There is therefore no extrinsic purpose to life; any purpose must be found within the development process itself. There are no absolute ethics, except perhaps that to resist the evolutionary process may be immoral and to contribute to it moral. This doctrine has become the basic dogma of secular humanism, the unofficial religion of the Western world.

This dogma has become so firmly grounded in our culture that respectable philosophers and scientists draw conclusions from it and do not shrink from advocating their conclusions no matter how strange they turn out to be. For example, Professor Crick, who shared the Nobel prize in medicine in 1962 for his contribution to the discovery of the structure of DNA, suggested several years ago[1] that it may be a good idea to redefine the concepts of "birth" and "death." He recommended that the time of birth of an infant be redefined as two days after parturition so that there would be time to examine it for defects. If its defects were sufficiently deleterious, the infant could presumably be eliminated with impunity because it had not yet become "alive." Similarly, Crick proposed redefining death as occurring when a predetermined age, such as eighty or eighty-five, is achieved. At that time the person automatically would be declared dead and all of his property would pass on to his heirs. Here, too, he could be eliminated with impunity, because he was already legally "dead."

Whatever else we might have to say about these suggestions, they are consistent with the doctrine of evolution, since they appear to lead to an improvement of the species in the long run.

Another example of a proposal drawn from evolutionary roots was put forward by Professor Bentley Glass about fifteen years ago.[2] He suggested a more natural definition of the concepts of "good" and "evil" whereby they would be completely divorced from their moral connotations. In fact, in his opinion the concepts of morals and ethics are illusory. His "natural" guidelines would define "good" as what is good for the development of the species; what is not good for the development of the species would, by definition, be "evil." He says, for example, that the eye is for seeing, and seeing is beneficial. Therefore should we not say that to see is "right" (or good) and not to see is "wrong" (or evil)? In this fashion he suggests the present moral basis of the Western world be exchanged for his more "natural" one.

## DARWIN'S CONTRIBUTION: RANDOM MUTATION AND NATURAL SELECTION

The doctrine of evolution did not begin with Darwin. It existed long before his time. References to similar ideas appear in the writings of the ancient Greeks. A generation or more before Darwin, the evolutionary doctrine was proposed by Buffon, Lamarck, Erasmus Darwin (Charles's grandfather), Geoffry St. Hilaire, and others. One might ask, then: If Darwin did not originate the evolutionary doctrine, what was so significant about his contribution that the world dates evolution from him?

Darwin's contribution was that he proposed a "theory" that bestowed on the doctrine of evolution a certain measure of reasonableness, which made it acceptable. He proposed the twofold process of random variation and natural selection. Darwin's variation is today interpreted as random mutations in the hereditary information store, which are preserved in future generations. Many such changes have occurred and the results of these changes presently exist in all populations of living organisms. If one of these changes should have a characteristic that gives to its host organism some kind of advantage in the struggle for survival, then that organism will produce more offspring than its fellows, and all its offspring will receive the same new beneficial feature and will in turn pass it on to their progeny. In this manner Darwin suggested that the form that possesses this hereditary advantage will eventually become the majority in the population. When this happens one could say that the population itself has changed. After a sufficient number of such changes the new population will eventually be recognizable as a new species. Moreover, Darwin suggested that this process continues to operate at higher and higher levels. Gradually the population not only gives rise to a new species, but it can even give rise to a new genus, a new family, a new order, a new class, and eventually to a new phylum. In this fashion Darwin suggested that life started from the simplest of forms, developed into a multicelled organism from which the invertebrates and then the vertebrates arose. Of the vertebrates, first there were the fish, then the amphibia, the reptiles, the birds, and the mammals. From a small carnivorous mammal the apes developed, and from the apes there arose man. Thus Darwin's theory explained the doctrine of evolution. Darwin's logic was so convincing that almost all scientists eventually accepted the theory, and with it they also accepted the doctrine.

The importance of the role of Darwin's theory in the acceptance of the doctrine of evolution cannot be overemphasized. Without Darwin's theory there is no explanation of the evolutionary doctrine. Darwin understood very well that the doctrine could not be accepted unless his theory were accepted.

The argument that the world came into existence by chance, and in particular that life developed by chance, did not originate in Darwin's generation or in his century. This argument was proposed long before his time, but it was not generally accepted because of the strength of the counterargument that the biological world is so well put together in such a complex manner that the probability of this complexity occurring by chance is small enough to be negligible.

## RABBI AKIBA'S PROOF

According to the *Midrash*, a disbeliever once approached Rabbi Akiba, and asked him, "Who created the world?"

"The world was created by the Lord blessed be He," answered Rabbi Akiba.

"Prove it," he countered.

"Who wove your garment?" asked Rabbi Akiba.

"A weaver, of course."

"Prove it," demanded Rabbi Akiba. He then turned to his students and said, "Just as the garment testifies to the weaver, the door to the carpenter, and the house to the builder, so the entire world testifies to the Holy One blessed be He Who created it."[3]

This argument can be understood by the following analogy. Suppose one leaves the earth via a spaceship and visits a distant planet where no one is known to have been before. Suppose further that upon disembarking from the spaceship and surveying the alien terrain, the traveler finds a book written in English, in good style, in a logical manner, on a serious subject. Would it be reasonable for the traveler to retain his original premise that no one has ever before visited this planet? Could he reasonably argue that this book—the paper, the binding, the ink on the pages, the formation of the ink on the pages into letters, words, and meaningful sentences—came into being by pure chance? Is it not obvious that this hypothesis is so improbable that the original premise that no one preceded him to this planet must be discarded? Because the chance hypothesis is so highly improbable he must conclude that the book was written by someone who knew what he was writing and that in some fashion the book was transported to this place.

Before Darwin's time only a negligible minority believed that life developed by pure chance, because the probability of such an event occurring by chance seems, to our intuition, to be so small as to make the event impossible. Then came Darwin and made his very important contribution. He proposed his dual process of random variation and natural selection that could vastly increase the probability of the chance development of life over what it had been previously thought to be. In this way the counterargument of the improbability of the chance occurrence of life, the argument cited above in the name of Rabbi Akiba, was apparently overcome. Darwin's proposal provided the means by which those who, for whatever reason, were dissatisfied or uncomfortable with the notion of a Divine Creator could mentally remove Him from the universe.

## THE NEED FOR A QUANTITATIVE CHECK

One must understand that Darwin's theory is entirely qualitative; it is not quantitative at all. Therefore, to say that Darwin proposed a mechanism that could vastly increase the probability of the steps necessary for the natural development of life is not really correct. The theory can only give the *impression* that it overcomes the counterargument of the chance development of life (as exemplified in the above *Midrash*). The fact is that the increased probability of chance development of living organisms implied by Darwin's theory requires quantitative checking to see if the resulting probabilities are indeed large enough to allow for evolution.

In the first hundred years since the publication of Darwin's theory such calculations were not performed, principally because the necessary knowledge was not available. That the doctrine of evolution was accepted on the basis of a nonquantitative theory is the shame of science. The theory that increased the extremely low probabilities of the events required for evolution was accepted without a quantitative investigation of the magnitudes of the resulting probabilities. The theory that seemed to be able to overcome the objection raised by Rabbi Akiba was received into the body of science without verifying that it in fact could overcome this objection.

In the last thirty years, however, we have begun to learn the secrets of heredity, and a quantitative examination of Darwin's theory is becoming possible. Between ten and fifteen years ago I published a series of papers that show that the probability of evolution by chance, even with the enhancement gained by Darwin's theory, is many times too small to explain evolution. Other physicists and mathematicians have also arrived at this same conclusion.

Nevertheless, evolutionists continue to hold fast to their doctrine, not understanding, or not wanting to understand, what the physicists and mathematicians are telling them. They continue to snow the opposition as they have always done by referring to the "vast" amount of evidence for the theory: the fossil record, embryological development, and vestigial organs.

## CYTOCHROME-C CONTRADICTION

Careful examination of this cited evidence shows (and I have shown this elsewhere)[4] that it is far from compelling, and in fact in a number of cases when a quantitative examination is made one can recognize that the evidence actually contradicts the Darwinian theory rather than supports it. In the past ten years or so evolutionists have begun to use the new biochemical results, particularly the amino acid sequences of various proteins. They claim that the studies of these sequences not only prove evolution but even yield numerical values for the ages of many categories of living organisms, by relating the number of amino acid differences between the cytochrome-C of two species to the elapsed time since their branching apart from a common ancestor.

About fifteen years ago I predicted that the new biochemical data that were then beginning to appear would lead to contradictions with evolutionary theory. At first, the effect of these data seemed to be otherwise, but after a number of years, contradictions did in fact begin to appear. The protein cytochrome-C, which plays an important role in the oxidation process of all living cells, has been extensively investigated. The amino acid sequence of this protein has been determined for many different species. According to evolutionary theory, the similarities between the corresponding sequences belonging to different species of organisms should follow the family relationships between them. One of the contradictions that results from these amino acid sequence studies is that the rattlesnake is more closely related to man than it is to any other species whose cytochrome-C sequence has been determined, including other reptiles. Furthermore, amino acid sequence studies have shown that the so-called divergence of the monkeys from the old-world monkeys is not consistent with the fossil record.

## FOSSIL-RECORD GAPS

Lately, because of the accumulation of data that appear to contradict the evolutionary theory of Darwin, particularly the stubborn refusal of the fossil-record gaps to close, a revolution seems to be occurring in evolutionary biology. A number of evolutionists are beginning to suggest that the evolutionary theory must be changed, that Darwin's theory is no longer adequate. So far no suitable substitute theory has been offered. Nevertheless, no one has even suggested that perhaps the evolutionary doctrine should be dropped, in spite of the fact that it was accepted only because of Darwin's theory.

## DOES MICROEVOLUTION
## CAUSE MACROEVOLUTION?

In 1980, a conference was held in Chicago on the subject of macroevolution. According to a published report on the conference,[5] the biologists present concurred that Darwinian evolution and its heir, the modern synthetic theory, are inadequate for explaining the facts. There seems to have been a distinct consensus that *microevolution*, the small adaptive changes that are observed to occur in nature, cannot be the cause of *macroevolution*, the supposed large evolutionary steps that are responsible for the development of new higher categories such as classes and phyla.

Darwin's key point, and the major thrust of the evolutionary theory that proceeded from his work, was that the development of the major categories, vertebrates from invertebrates, amphibia from fish, mammals from reptiles, and so forth, was merely an extension of the microevolutionary processes that can be observed in nature. There is now a strong movement in the field of biology that denies this hypothesis.

No one, however, even hints or suggests that the evolutionary doctrine be rejected. The doctrine has apparently taken on a life of its own and has become its own *raison d'être*. The scientists of our generation seem to be capable of rejecting the theory that initially made the doctrine plausible and acceptable while continuing to hold on to the doctrine itself. Our sages tell us that if one strives to be a believer, then God will help him achieve his goal. If, on the other hand, he strives to be a disbeliever, then he will also be helped to achieve this goal.

## BRIBED INTO BELIEF

The question arises (and I have been asked this question many times): If the evidence for evolution is as weak as I have indicated it to be, then why do the majority of the biologists, most of them educated and intelligent people, still hold fast to the evolutionary doctrine and profess the belief that the biological world came into being by chance? This question was addressed by one of the sages of the previous generation, Rabbi Elchanan Wasserman. He cited the opinion that the first verse of the Ten Commandments, "I am the Lord your God . . . ," is itself one of the Commandments. He then asks how, according to this opinion, one goes about observing this Commandment. The Commandment is to believe in God. Is not belief or disbelief the starting point of observance or the lack thereof? If one does not believe, then one does not even approach this Commandment with a serious intent. If, on the other hand, one does believe, then how can one perform any action to observe this Commandment since one already fulfills it? Moreover, if we accept the statement of Rabbi Akiba that the entire world testifies to the Lord, blessed be He, Who created it, then why should we need such a Commandment as this one? Should it not be obvious even to a child that there is a Creator of the universe? One should be expected to believe such a self-evident fact without the need of a special commandment. On the other hand, however, what should be obvious to even a child seems to be far from obvious to a large number of educated and intelligent philosophers and scientists. How, Rabbi Wasserman asks, can one explain this phenomenon?

To answer his question, Rabbi Wasserman points to the negative Commandment forbidding judges to take bribes. He notes that according to *halakhah*, the minimum value that is considered a bribe is one *perutah*, which is the equivalent of about one U.S. penny. According to the *halakhah*, if a judge receives even one *perutah* from a litigant in a court case, the judge is disqualified from sitting on that case. Suppose that the judge were a great man, a wealthy man, and an extremely honest and fair man. Would it be reasonable to suppose that if a litigant pressed a *perutah* into his possession, that this little *perutah* could have any influence at all on his judgment in the case? Nevertheless, the *halakhah* disqualifies the judge from sitting on the case. He is disqualified because this little *perutah* can have an influence, perhaps only subliminal, on even a great and honest judge. The effect may be very subtle, but even the smallest bribe can have some effect on him and therefore he is disqualified.

The situation is similar to that of the scientist or philosopher who holds the belief that life arose and developed by chance alone. This belief provides comfort to him and to others who do not want to be subject to the Ruler of the Universe. If life arose by chance, then there is no Creator. If there is no Creator, then there is no Divine Ruler. He believes himself free. He has committed himself and his reputation to this belief. Moreover, rejection of the evolutionary doctrine by a biologist today is decidedly unfashionable. To come out strongly against the evolutionary doctrine could cost him dearly in the influence he commands in his field. Thus his vested interest in the evolutionary doctrine and his apparent need to deny the Creator play the role of the judge's bribe in preventing him from adopting a truly unbiased view of the question of evolution. If the idea should ever arise in his mind to question the doctrine of evolution, he will drive it from his thoughts because of the uncomfortable consequences that might follow if he were to arrive at the "wrong" conclusion.

The average intellectual, and particularly the student of biology, is still being intimidated into accepting the doctrine of evolution. During his student days he sits as part of the captive audience of a professor who has his own need to preach the evolutionary doctrine. The discriminating student should recognize that the doctrine lacks a proper scientific basis and he should refuse to be intimidated. Those desiring real truth must seek out a proper responsible teacher of Torah who will be able to lead them, without apologetics to pseudoscience, into the framework of *daat* (knowledge of) *Torah*.

From *B'Or Ha'Torah* 2, 1982.

# 14

# Information Theory Considerations of Organic Evolution

## Lee M. Spetner

According to Darwinian theory, a population of living organisms adapts to its environment through the dual process of random mutation and natural selection. If the population adapts to the environment in the sense that its hereditary genetic makeup changes to a configuration that causes a better-adapted organism to develop, then in a very general sense, information has been transmitted from the environment and received into the genes of the members of the population.

The probability of the occurrence of chance genetic changes can be used to quantify the information that is gained by the genes in an evolutionary step. This information is situated in the sequence of nucleotides that make up molecules of the gene. We let $I$ denote the amount of information required in order to effect an adaptive advance in one evolutionary step. We do not know the value of $I$ because we do not have the vast knowledge necessary to tell, for any particular environment, which new configurations of DNA nucleotides will re-

179

sult in an organism better adapted to the environment. If, however, we use a reasonable model of independence of the nucleotides and uniform probability of point mutations, we can calculate the probability of actually transmitting bits of information. In an evolutionary step consisting of a sufficiently large number of trials (or births), that is, anywhere from $10^{12}$ to $10^{14}$, this probability turns out to be approximately

$$P = 1 - (1 - 2^{-1})^{3n} \qquad (1)$$

where $n$ is the number of nucleotides in the DNA that are available for change.[1]

Although we do not know the value of $I$, we note from (1) that as we make $I$ larger, $P$ gets smaller and vice versa. Thus we see that the larger the amount of information required to advance one evolutionary step, the smaller is the probability of achieving it. We can overcome the difficulty of not knowing the magnitude of $I$ by dealing with the average information, $<I>$, transmitted per step, defined as the product $PI$.

While we are interested in the adaptive information that is received into the genes, we must also note that the genes contain a great deal of information that pertains to their own functioning in what is understood to be their primary role, directing the manufacture of protein at the cellular level. This type of information is analogous to grammatical structure in information-bearing texts. We shall therefore consider the information developed by the evolutionary process to be composed of two parts—adaptive information, which we denote by $i$, and grammatical information, which we denote by $j$—so that the total information $I$ can be expressed as

$$I = i + j \qquad (2)$$

The amount of grammatical information in any message in a natural language is some fraction of the message length. The amount of this grammatical information, usually referred to as redundancy, has been determined to be about 82% for English (a high-redundancy language) and about 17% for Hebrew (a low-redundancy language).[2] We can reasonably say that the DNA message similarly has grammatical structure, or redundancy, that is some fraction of the message length. We shall let this fraction be denoted by $B$. Then the grammatical information is (in bits) $2Bn$, where $n$ is the length of the DNA sequence to be modified, each of whose symbols is worth 2 bits of information.

The average adaptive information is then given by

$$<I> = (I-2Bn)P \tag{3}$$

where $P$ is given by (1). In Figure 14–1, the average information $<I>$ is plotted as a function of the information $I$ that must be transmitted from the environment to the species in order to achieve a step of evolutionary advance. A family of such curves is shown for various values of $n$, the sequence length available for change. The fraction of required grammatical information is here taken to be 0.1. Note that each curve rises to a maximum as $I$ is further increased. This fall-off reflects the fact that when the information required for an adaptive advance is large, then the probability of achieving it is very small, so that the average information (the product of the information and the probability) is small. Thus, for each value of $n$ there is a maximum amount of information that can be incorporated into the gene in one evolutionary step. From the figure we can see that for $n = 2$ the maximum

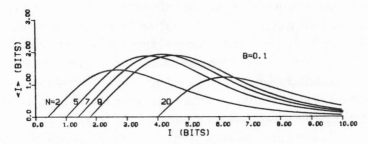

Figure 14–1

information per evolutionary step is about 1½ bits. As $n$ increases from 2, this maximum value rises. Note that this maximum rises to a peak value of about 1.9 bits for $n = 7$ and then falls off for larger values of $n$. From this figure we see that for $B = 0.1$ the maximum information transfer that can be achieved under the best possible conditions is less than 2 bits.

In order to gain some idea of the amount of information that resides in DNA, and which the evolutionists must claim has been accumulated through the random processes of mutation and natural selection, we note that in a mammal there are typically about $10^{10}$ nucleotides in the DNA of each cell. If all of these nucleotides represented information, there would be about $2 \times 10^{10}$ bits of genetic information in a mammal. Some of this information may be claimed to be either nonfunctional or copies of the same thing one or more times, although we do not really know. Even if we should say that 90% of these nucleotides do not represent real information, there would still be $2 \times 10^9$ bits of information in the genes. The question then arises: How did this information get there? To say that it found its way there through the process of random mutation and natural selection would require orders of magnitude more time than is available under even the most generous estimate of the age of the Earth. This is because at 2 bits per evolutionary step, we would need about $10^9$ evolutionary steps, each one of which requires somewhere from $10^4$ to $10^6$ years.

If the requirements on the grammatical structure were relaxed and dropped from 10% to 1%, then the maximum amount of information transfer per evolutionary step would occur for $n = 70$ and would amount to about 4.6 bits. The required number of evolutionary steps would then drop from $2 \times 10^9$ to about 450 million, still requiring orders of magnitude more time than could possibly have been available.

One possible, but not very good, answer can be given by the evolutionists to their dilemma. They could say since we really do not know for sure how much real information is represented by the large amount of DNA in the genes, perhaps it is really many orders of magnitude smaller than our small estimate of only 10% of the apparent information. This might be the case if for each evolutionary step there were a very large number of different adaptive forms possible from which the random process could choose. If the real information is sufficiently small, then there might be enough time for it to accumulate through the evolutionary process of random mutation and natural selection. This would mean that at each evolutionary step an enormous number of adaptive configurations would be possible. If this were the case, however, the concepts of convergent and parallel evolution would not be possible, causing the entire phylogenetic classification system to collapse.[3] In consequence, evolutionary theory would not be able to explain the related structures of living forms since the current explanation relies heavily on the concepts of convergent and parallel evolution.

In conclusion, the above considerations show that the process of random mutation/natural selection is not able to represent a suitable mechanism for explaining the evolutionary doctrine.

From *B'Or Ha'Torah* 2, 1982.

# 15

# A Statistician Looks
# at Neo-Darwinism

## Avraham M. Hasofer

### THE THEORY OF EVOLUTION DOES NOT HAVE A
### SHRED OF EVIDENCE TO SUPPORT IT

In his "Letter on Science and Judaism" of 1962[1] the Lubavitcher Rebbe *shlita* writes: "If you are still troubled by the theory of evolution, I can tell you without fear of contradiction that it has not a shred of evidence to support it."

Such a statement comes as a shock to those of us who have been indoctrinated since our school days with the notion that evolution is a scientific fact. Most of us retain that notion throughout our life, and it is only the few who become involved in advanced study and research in evolutionary theory who have the opportunity to examine the foundations on which it is built.

Having been involved for several years in that research, I feel it my duty to come out and share my conclusions with the public at large, at a time when much confusion exists about this subject. A substantial part of what I would like to say has already been covered

185

by Dr. L. M. Spetner's excellent articles which appeared in *B'Or Ha'Torah* 2 and to which I refer the reader. The remarks below are to be taken as complementary to his.

## WHAT IS THE THEORY OF EVOLUTION?

The theory of evolution discussed in modern biology books invariably means neo-Darwinist theory, which purports to explain the appearance of the diverse life-forms existing on Earth. According to this theory, plants and animals have descended from simple ancestors far back in the history of the Earth. The mechanism by which new species appear consists of two steps:

1. the appearance of new features through mutations
2. natural selection, which ensures that beneficial mutations will survive and be passed on to the offspring.

## WHAT ARE THE CRITERIA FOR
## A SCIENTIFIC EXPLANATION?

The question arises: Does neo-Darwinism satisfy the accepted criteria of a scientific explanation?

There are many scientific theories which purport to explain observable facts. Let us consider a typical example: the perturbations of the orbit of the planet Uranus. It was suggested by the astronomers Adams and Leverrier that the perturbations were due to the attraction of a hitherto unobserved planet. Using Newtonian mechanics, they proceeded to calculate the expected position of the unknown planet. The latter was then observed by Galle in 1846, at the very position which had been predicted by calculation.

This example illustrates two basic criteria for a scientific explanation:

1. The proposed cause must be observed to exist.
2. It must be possible to show *quantitatively* that the proposed cause explains the observed effect through the use of accepted theory.

Observation of the cause and the effect then provides a confirmation of the theory used in the calculation as well. Thus the observation of Neptune provided confirmation of Newtonian mechanics, on which the calculations were based.

Sometimes, however, the proposed cause is unobservable, as in the case of the electronic theory of electricity. In that case, it is absolutely imperative that the cause should quantitatively predict the effect through the use of accepted theory.

## DOES NEO-DARWINISM SATISFY THE CRITERIA?

The answer to this question is emphatically "no." In fact, the proposed causes have never been observed:

1. A new species has never been observed descending from another.
2. A crucial aspect of the proposed mechanism of evolution is the appearance of mutations which are beneficial in some environments. However, a beneficial mutation has never been observed. This fact is usually well camouflaged in texts which support evolution. Examples of selection of a particular advantageous feature abound. But close examination of the facts invariably reveals that the advantageous feature could well have been there all the time. There is no evidence that a mutation has occurred.

One very popular case study quoted by Darwinists is that of the peppered moth (*Biston betularia*) in Great Britain. In the past, there were many light-colored moths that blended with the lichen growing on trees. As industry spread, the lichen disappeared, and the dark (melanic) form of the moth became predominant. This predominance has been attributed to the adaptive advantage of the dark color, which provides the moth with better camouflage against predators when it settles on soot-blackened tree bark.

The operation of natural selection in this case has been amply established by observation and experiment. But it has never been established that the melanic form of the moth appeared through mutation. The more cautious writers such as Ruse[2] do not make such a claim. But in the *Encyclopedia of Nature and Science*,[3] under the article "Evolution," we read: "Until about 100 years ago, nearly all the peppered moths were white. . . . Occasional black individuals appeared as the result of a mutation. . . ."

In fact, the most probable explanation of the origin of the dark variety is that it had existed all the time but only as a very small proportion of the population, possibly because the allele responsible for the black color also offered some other selective advantage. Such a situation is referred to in the literature as genetic polymorphism and is very widespread.[4]

One situation where the mutation rate was artificially increased is the Bikini atoll after the atomic bomb tests. Large numbers of mutations were observed in the fauna and the flora, all invariably deleterious.

It is to be emphasized that cases of natural selection abound. But even the most extreme evolutionists have not claimed that natural selection without mutations has led to the appearance of new species. The only way to redeem the mutation aspect of neo-Darwinism is to show by calculation that, although beneficial mutations have not been observed, they *could* have occurred during geological times.

## THE PROBABILITY OF BENEFICIAL MUTATIONS

Until about twenty years ago, it was not possible to perform any numerical calculations in relation to the occurrence of beneficial mutations, because the mechanism of occurrence of mutations was not well enough understood.

With the discovery of the chemical composition of chromosomes, it has become possible to put numerical values on the probability of appearance of a beneficial mutation. To my knowledge, the first time such a calculation was published was in L. M. Spetner's paper in the *Journal of Theoretical Biology* in 1964.[5] A simplified version was subsequently published by myself.[6]

Spetner interpreted the results of the calculations in terms of speed of transmission of information from the environment to the living organism. This interpretation was further extended by him in the

*IEEE Transaction on Information Theory* in 1968.[7] It was reiterated in a paper published two years later in *Nature*.[8]

Briefly, the results of the calculation indicated that the probability that even one beneficial mutation would have occurred within the most generous allocation of geological time is so low that it is unreasonable to assume that it had occurred. Further details of the calculation will be given later on.

As has been mentioned above, Spetner's calculation has been available in published form since 1964. In view of the central importance of this problem to evolutionary theory, one would have expected a flurry of attempts to produce alternative calculations which would make the situation more palatable. But there have been no attempts at all! A few weeks ago I made a thorough search of the *Science Citation Index* to find out whether any answers to Spetner had been published. There were none.

My own contacts and discussions with biologists have confirmed the fact that no plausible calculation has been developed to support the claim that the beneficial mutations required to sustain evolution can occur with a reasonably high probability.

## SOME DETAILS OF THE CALCULATIONS

To obtain an idea of the order of magnitude involved, we visualize the chromosomes of a living organism as made up of a string of nucleotides. A reasonable total length is about one billion nucleotides, each capable of having one of the four bases: adenine, guanine, cytosine, and thymine. The molecular mutation rate is taken to be $3 \times 10^{-9}$ per birth per nucleotide, and we assume that mutation of one base to any one of the other three bases is equiprobable. The length of a particular locus in the chromosomes is taken to be 15 nucleotides. The total number of possible sequences is $4^{15} = 10^9$.

One can then calculate the mean number of births required for a beneficial mutation to occur. This depends of course on the number of nucleotide sequences which would be beneficial in a particular environment. We thus obtain the following values:

| Number of beneficial sequences | Mean number of births required |
|:---:|:---:|
| 10,000 | $0.18 \times 10^{45}$ |
| 100,000 | $0.30 \times 10^{39}$ |

Let us now consider the required birthrate. First we note that the most generous of geologists will allow as an upper bound of the age of the Earth six billion years ($6 \times 10^9$).[9] Even if we allow the wide choice of one hundred thousand beneficial mutations, and the whole span of six billion years for the one beneficial mutation to occur, we would still expect to need a yearly birthrate of $5 \times 10^{28}$. This is certainly totally unrealistic for any higher living organism on Earth.

In fact the total area of the Earth is $5 \times 10^{14}$ m². Suppose one higher organism were born each year on each square meter of the Earth's surface for six billion years. Then we would still have only $3 \times 10^{24}$ births, not even one ten-thousandth of the average number of births for one beneficial mutation to occur! (For further technical details of this calculation, see Appendix A.)

One can recast the above results within the framework of the universally accepted Neyman-Pearson Hypothesis Testing Procedure.[10] Suppose that we still allow a choice of one hundred thousand beneficial mutations, and let us assume a population with one million births per year. The hypothesis to be tested is: "At least one beneficial mutation has occurred during the time span available to the population." If we choose a significance level of 5%, then since this probability corresponds to a number of births less than or equal to $0.7 \times 10^{18}$, it follows that the hypothesis must be rejected if the time span available is less than $0.7 \times 10^{12}$ years. Since geological data allow only $6 \times 10^9$ years as an upper bound, the hypothesis must be rejected.

Another way of looking at the problem is through the technique of Bayesian inference,[11] which allows us to update our degrees of belief in scientific explanations in the light of new knowledge. Suppose that initially, before the chemical structure of DNA was discovered, a biologist was 99% sure that neo-Darwinism accounted for the appearance of the diverse forms of life. After the biologist becomes aware of the calculations presented above, he realizes that the probability that beneficial mutations have occurred at all during geological times is far less than 0.0048. If he behaves rationally, he must then conclude that the chance that neo-Darwinism accounts for the diversity of life is far less than 32.2%. In other words, he must now be much more than 67.8% sure that the diversity of life is due to other, hitherto undiscovered causes. (For further details of the Neyman-Pearson and Bayesian procedures, see Appendix B.)

Clearly the above calculations heavily understate the case against the neo-Darwinist mechanism, since the latter requires not one beneficial mutation but a very large number—probably billions. But since even *one* mutation appears improbable, so much the more a large number!

## Discussion

The calculations performed in the preceding section are based on our present understanding of the chemical coding of genetic information in the chromosomes. As explained in this chapter and pointed out by Salisbury,[12] this understanding is quite incompatible with the concept of evolution by natural selection of adaptive genes that are originally produced by random mutations.

It has already been mentioned that:

1. A new species has never been observed descending from another.
2. The appearance of a beneficial mutation has never been observed.

In addition, calculations based on our present understanding of the genetic code show that the probability of the appearance of beneficial mutations in geological times is infinitesimally small. The only rational conclusion that can be drawn from all of the above is that neo-Darwinism is a speculation unsupported by fact.

It is up to the evolutionists to bolster up their case by supporting it with facts and/or *quantitative* calculations. Until they do, one must just state that, at this stage, science cannot shed any light on the mechanism of appearance of the diverse forms of life on Earth because the only proper scientific attitude in the absence of evidence is to withhold judgment.

Evolutionists are well aware of the fatal weakness of neo-Darwinism in the area of mutations. Some attempt to evade it through a naive faith. For example, in the 1974 volume of the *Encyclopaedia of Nature and Science*,[13] in the article "Evolution," we read: "Mistakes, or mutations, sometimes occur when the (genetic) instructions are being handed down to the offspring. These wrong instructions are generally harmful and the organism gains no benefit—it may even die—but the mutations sometimes result in major improvements.... Mutations of this kind *must* have happened many times during the long course of evolution...."

Other evolutionists use purely verbal arguments, too weak to stand the test of being reduced to real numbers. For example, Ruse[14] writes:

> Consider a monkey, sitting at a typewriter, randomly striking the keys. *Prima facie*, the production of life by random processes seems about as likely as the monkey's typing out the whole of *Hamlet*, ... It may not be logically impossible; but ... it is practically impossible. Suppose, however, that every time the monkey strikes the "right" letter, it records; but suppose also that "wrong" letters get rubbed out [literally or metaphorically!]. And suppose that elimination of the wrong letter is the full consequence of a "mistake": one does not lose what has already been typed.... The typing of *Hamlet* no longer seems like anything so impossible, even by a "blind law" phenomenon, like a typing monkey....

Such verbal arguments may sound plausible to a nonmathematician, but for anyone who has had experience with the huge numbers generated by combinatorial arguments, nothing short of an actual calculation will carry any weight. And all calculations made so far emphatically deny the likelihood of beneficial mutations appearing within geological times.

# WHY IS NEO-DARWINISM SO POPULAR?

Spetner[15] asks: "If the evidence for evolution is as weak as I have indicated it to be, then why do the majority of biologists, most of them educated and intelligent people, still hold fast to the evolutionary doctrine . . . ?" His answer, if I understand it correctly, is in two parts:

1. Evolutionary doctrine provides comfort to those who do not want to be subject to the Ruler of the Universe.
2. Biology students are intimidated by the threat of ostracism into accepting neo-Darwinism because of the large interests invested in the current view.

While these two factors are undoubtedly important, I would like to bring in a third one, which I believe to be even more fundamental, since it has its source in a basic psychological drive. This factor is highlighted by the Lubavitcher Rebbe *shlita*[16] in his previously quoted letter:

> This question may be asked: If the theories attempting to explain the origin and age of the world are so weak, how could they have been advanced in the first place? The answer is simple. It is a matter of human nature to seek an explanation for everything in the environment, and any theory, however farfetched, is better than none, at least until a more feasible explanation can be devised.

## APPENDIX A

In this appendix, the formula used to calculate the probability distribution of the number of births required to achieve one beneficial mutation is given. This will be followed by a discussion of the parameter values adopted.

The derivation of the formula (in a trivially different form) is given in my simplified treatment of Spetner's natural selection model in the *Journal of Theoretical Biology*.[17]

Let

$N$ = number of births until a beneficial mutation is reached.

$n$ = number of nucleotides forming a beneficial sequence. Take $n$ = 15.

$p$ = probability of mutation of one nucleotide. Take $p = 3 \times 10^{-9}$.

$M$ = total number of possible sequences of length $n = 4^n$. If $n=15$, $M = 10^9$.

$m$ = number of beneficial sequences out of the $M$ possible ones.

$P(N)$ = probability of mutating into a beneficial sequence in $N$ births or less.

Put

$$s_k = (p/3)^k, \quad k = 0, \ldots, n.$$

$$q_k = \binom{n}{k} 3^k / M, \quad k = 0, \ldots, n.$$

Then

$$P(N) = 1 - \left\{ \sum_{k=0}^{n} q_k \exp\left(-N s_k\right) \right\}^m.$$

In the Neyman-Pearson test the following values are used: $n = 15$, $m = 100{,}000$, $p = 3 \times 10^{-9}$, $N = 0.7 \times 10^{18}$. It then turns out that $P(N) = 0.05$ (5%).

If on the other hand, we put $N = 6 \times 10^{15}$, representing a yearly birthrate of one million over a period of $6 \times 10^9$ years, then $P(N) = 0.0048$.

The expected number of births until a beneficial mutation is obtained is calculated from $P(N)$ by numerical summation.

The value $n = 15$ for the length of the nucleotide string is based on the generally accepted fact that about three to seven amino acids are essential for the functioning of an enzyme. Since 3 nucleotides are needed to code one amino acid, the essential part of the enzyme will be coded by 10–20 nucleotides. The number 15 is taken as an average value.[18]

As far as the value $p = 3 \times 10^{-9}$ is concerned, it is near to the value calculated by Champe and Benzer,[19] namely, $10^{-8}$, for T4 phage.

It is interesting to note that too high a rate of nucleotide mutation will, in fact, reduce the probability of a beneficial mutation appearing, because at the same time that the beneficial mutation must appear, the rest of the information coded in the chromosomes should remain undisturbed. Otherwise, since the immense majority of mutations are deleterious, and many are lethal, the organism would not survive.

Spetner (1964)[20] has made a rough calculation of the optimal mutation rate.

If : $\lambda$ = number of nucleotides representing essential information without which the organism cannot survive, then the optimal value of $p$ is approximately $n/\lambda$.

Now the total number of nucleotides varies from $10^7$ to $10^{11}$.[21] Assuming $\lambda = 10^{10}$, then the optimal value of $p$ is $1.5 \times 10^{-9}$, a value not far from the one adopted.

Finally, as far as the value $m = 100,000$ for the number of beneficial mutations occurring, it is to be pointed out that since the total number of possible sequences of 15 nucleotides is $10^9$, the above value means that one in 10,000 mutations is assumed to be beneficial, a very generous assumption in view of the fact that a beneficial mutation has never been observed to appear.

# APPENDIX B

### Neyman-Pearson Test of Hypothesis

The distribution function of the number of years required for a population with one million births per year to produce a beneficial mutation is known (see Appendix A).

Choosing a level of significance $\alpha = 5\%$, which is the usual value in biometry, we set up a "critical region" $0 < N < N_0$ such that $P(N_0) = \alpha$. As stated in Appendix A, if $\alpha = 5\%$, $N_0 = 0.7 \times 10^{12}$. If the geological age of the Earth falls within the critical region, then the hypothesis, namely, "at least one beneficial mutation has occurred in a population with one million births per year during the geological age of the Earth," must be rejected.

**Bayesian Inference**

Let $B$ stand for the event that life exists on Earth as we see it, following the laws of genetics.

Let $A$ stand for the neo-Darwinist theory.

Let $A^c$ stand for all other possible explanations of the appearance of life on Earth (including those as yet undiscovered).

Let $P(A)$ stand for the probability of $A$ (in the sense of degree of belief).

Suppose $P(A) = 0.99$. Then $P(A^c) = 0.01$.

Let $P(B/A)$ be the likelihood that life has appeared on Earth in accordance with the neo-Darwinist theory.

From the calculations given in Appendix A, it appears that $P(B/A) \ll 0.0048$.

We let $P(B/A^c) = 1$, which means that there does exist a fully satisfactory explanation (possibly as yet undiscovered) of the appearance of life on Earth. This reflects our basic belief in the power of science.

Then Bayes's theorem states

$$P(B/A) = \frac{P(B/A)\ P(A)}{P(B/A)\ P(A) + P(B/A^c)\ P(A^c)}$$

$$\ll \frac{0.0048 \times 0.99}{(0.0048 \times 0.99) + (1 \times 0.01)}$$

$$= 0.322$$

Thus, even if the a priori probability of neo-Darwinism is 0.99, the a posteriori probability, after realizing the improbability of beneficial mutations occurring, is less than 32.2%.

From B'Or Ha'Torah 3, 1983.

# 16
# Torah, Science, and Carbon 14

## Yaacov Hanoka

The Lord made one thing opposite to another
—Ecclesiastes 7:14.

The method of radiocarbon dating is based on a number of assumptions which raise unanswerable questions like: Is the ratio of carbon 14 to carbon 12 a constant? Is the atmosphere the source of carbon dioxide for all living organisms? Is the intensity of cosmic rays striking the Earth uniform? Has the $C^{14}$–$C^{12}$ ratio been reduced by the industrial revolution? The extrapolation process used to configure carbon 14 dating is not strictly speaking good science, whereas the more reliable, empirical method of tree ring dating gives accurate results going back only 5700 years.

For many people, the difficulties of reconciling science and Torah revolve around the question of dates. The Torah asserts that the world is 5,742 years old. Science claims it is something on the order of several billion years old. There seems to be a slight discrepancy here. But is there really? Let us look at one aspect of this question by investigating

197

one of the more reliable of the dating techniques used in science, that of the carbon 14 dating method (also known as radiocarbon dating).

## THE CARBON 14 DATING METHOD

Carbon as we ordinarily know it is referred to as carbon 12. The "12" indicates the total number of neutrons and protons which make up the nucleus of the carbon atom. There is another form or isotope of carbon which has two additional neutrons in the nucleus, and this is called carbon 14. Carbon 14 is radioactive and decays at a known rate, such that in a time span of about 5,700 years half the original amount will be left, the other half having undergone radioactive decay.

In the outer atmosphere surrounding the Earth, carbon 14 is formed through a nuclear reaction. The energy to foster such a reaction is supplied by the cosmic rays which are a form of very energetic radiation continually bombarding the Earth. These cosmic rays form energetic neutrons which then react with nitrogen 14 in the atmosphere to form carbon 14. This carbon 14, in turn, reacts with oxygen to form carbon dioxide.

The total amount of radioactive carbon in carbon dioxide as compared to ordinary carbon, carbon 12, is very small—less than 1%. This radioactive carbon dioxide along with ordinary carbon dioxide then distributes itself throughout the Earth's atmosphere and eventually is absorbed by plants and animals while they are alive. When a plant or animal dies, it no longer takes in any carbon dioxide. Thus, by measuring the amount of carbon 14 relative to the amount of carbon 12 in a formerly living thing, and knowing the decay rate of carbon 14, one can, by *extrapolating* back in time, calculate how many years ago this particular living thing died. This is the basis of the carbon 14 dating technique, briefly stated. Let us examine some of the assumptions underlying the method.

### Basic Assumptions and Their Variations

The basic premise supporting the entire method is that the ratio of carbon 14 to carbon 12 is constant and has not changed over the period for which any particular object is being dated. Implicit in this assumption is that the rate of carbon 14 production is the same everywhere on Earth and has always been so, and that the cosmic ray intensity striking the Earth is uniform all over the Earth at all times. A further assumption is that the source of carbon dioxide for living things is the atmosphere.

When the carbon 14 dating method was first suggested, its discoverer, W. F. Libby, was himself cautious about the evidence supporting the basic assumption.[1] Since that time, a considerable body of evidence has been gathered to indicate that, in fact, there have been variations and this first basic assumption must be tempered with these observations.[2] In addition, variations in the rate of carbon 14 production as a function of climate, season, and position on Earth have been found and analyzed extensively.[3] Furthermore, it has been learned that changes in the carbon 14 level can be brought about by changes in the cosmic radiation caused by changes in the Earth's magnetic field.[4] In the latter case, a number of significant changes, such as the reversal of the Earth's magnetic poles, have now been discovered, and these have occurred in recent history.

It is important to point out that the discovery of all of the above variations has not put into question, at least in the minds of most scientists, the basic validity of this dating method. Instead, it has induced a greater level of care in extrapolating the results and a greater sensitivity to the question of variations. In fact, things have almost become inverted in that there is now a considerable amount of research activity in using the carbon 14 method to study such variations. Also, it should be pointed out that two events in very recent history have also had some effect on the $C^{14}$–$C^{12}$ ratio. One such case is the industrial era over the last 100 years or so and the subsequent burning of fossil fuels on a large scale. This has produced a large amount of carbon dioxide in the atmosphere and a reduction in the $C^{14}$–$C^{12}$ ratio.[5] The second is the explosion of nuclear devices which has increased the amount of carbon 14 in the atmosphere quite significantly—as much as 40% in the Northern Hemisphere in the year 1962–1963 alone.[6] Finally, it should be mentioned that some kinds of plants which grew near areas which were rich in limestone were known to obtain some of their carbon 12 from the soil and not from the atmosphere, and thus they would have a somewhat different $C^{14}$–$C^{12}$ ratio than a similar plant which was not grown in soil rich in limestone.[7]

## TREE RING CHECKING

A desirable way to check on dates arrived at by radiocarbon dating is through comparison with dates arrived at by other methods. One method in particular which has received a good deal of attention and has been the subject of much research is that of dating by counting tree rings. Samples of the Bristlecone pine found in the western United States, for example, have shown tree rings which indicate ages going back several thousand years. The carbon 14 content of individual rings can then be measured and a radiocarbon date for the rings can be compared to the date arrived at by counting. The counting should be done under the microscope, as these rings are very closely spaced and not easily counted.[8] Since the validity of tree ring dating is accepted almost axiomatically, this was viewed as an excellent test of the carbon 14 dating method.

Agreement has been good, but not perfect, between these two dating methods.[9] Two things should be noted here. First, the oldest such tree ring specimen seems to be about 8,000 years old. Truly accurate tree ring measurements, however, date back only 5,700 years. So, the best validation of the carbon 14 method vis-à-vis the tree ring dating is based only on data going back 5,700 years—a figure interestingly close to the Torah age for the world: 5,742 years. Thus, assumptions of the validity of carbon 14 dating well beyond such a date cannot rely on the cross-check afforded by tree ring dating. In other words, extrapolating beyond this time period of about 6,000 to 8,000 years involves assumptions which are not nearly as readily cross-checked.

## EXTRAPOLATION

Extrapolation of assumptions into spatial areas or time periods beyond the reach of the original observation of the data is a basic (and problematic) principle used in science. The theory of evolution and certain theories of cosmology represent far more flagrant examples of this practice than does our case at hand. In our discussion of carbon 14 dating we see that science, *if it is good science*, is self-rectifying and ready to question and examine the validity of extrapolated conclusions.

The method of extrapolation is based on the tacit assumption that *all natural phenomena are uniform*.

The modern view of the physical world strongly suggests that natural phenomena occur in such a uniform way that the relationships which we discover governing physical phenomena are considered to be true not only under the directly observed circumstances but for as long or as far as we care to imagine or project.

A religiously oriented reader might say that this is a confirmation that God created the world in a way describable by simple mathematical formulae. Indeed, Jews bless God for having created the world with uniformity and regularity. But the believing Jew also has faith in the Creator Who provided us with unique, rationally unprovable events and phenomena.

## THE TORAH PERSPECTIVE ON SINGULARITIES IN THE EARTH'S HISTORY

What does the Torah say about extrapolating back into time? Can the general assumption of the uniformity of natural phenomena be found in the Torah? The example of carbon 14 dating shows that good science is and should be self-corrective. The argument was from science itself and its own findings. Let us now consider the Torah view on similar matters. We are not suggesting that we should use the Torah here to prove or disprove scientific findings, as this is not the primary function of Torah.

### Creation—The Greatest Singularity

The greatest singularity described by the Torah is, of course, Creation. The Torah tells us that God created a full-blown world from nothing. Man and the physical world were created in a complete way such that the first man was created as an adult and the physical world was all there and ready to be used by him.

Suppose a scientist had come along with his instruments about 100 years after Adam, the first man, was created. Suppose he tried to date decayed pieces of wood using the carbon 14 method. And suppose he had never heard that the world had been created 100 years before. Would he have been able to discover that in fact the world is only 100 years old? Or would he, by assuming that the world had not been created 100 years before and extrapolating back as far as he pleased, have come up with a different number for the age of a particular object, say, 2,000 years? If this object had contained a certain amount of Carbon 14 when created, then it is quite possible that our hypothetical scientist would have concluded, by indiscriminate use of the method of extrapolation, that this object was 2,000 years old. Of course, if he were a careful scientist, he would probably have discovered that there was some kind of unusual behavior in the carbon 14 formation rate just 100 years back—a singularity that would lead him to temper his initial conclusion with caution.

### The Singular Moment of Transformation of Adam and Eve

In addition to the act of Creation itself, the Torah gives another very significant example of a change in the physical world. This was the change which followed the sin of eating the forbidden fruit in the Garden of Eden. Prior to this event, man was immortal; he was to live forever in complete harmony with the world which God created for him. All his needs were provided him without any effort on his part. He did not have to till, plough, weed, harvest, or worry about food; it simply grew and was always available. After the sin, as is well known, man had to work very hard to wrest food from the soil, indicating a very dramatic change in the nature of physical things, i.e., a singularity, a very significant variation in the way the physical world was structured and governed. Since God, at this time, changed the nature of man and also the physical world, it is not unreasonable to assume that the $C^{14}$–$C^{12}$ ratio might have changed as well.

## The Sinai Singularity

Another major singularity was the composite of events surrounding the giving of the Torah at Mount Sinai. Here again, the condition of mortality for all the people present at this event was lifted. Until the sin of the golden calf, the people were immortal. Unfortunately, the sin of the golden calf caused the removal of this state of immortality, which nevertheless did exist, albeit for a rather short duration. There were other interesting aspects surrounding this entire episode. In fact, we are told that the sun stood still five times and that at each of these times, the day was three times longer than its normal length.[10] These five times were the Exodus itself, the splitting of *Yam Suf* (the Sea of Reeds, or the Red Sea), the battle with Amalek, the Giving of the Torah, and the battle with Sihon at the brook of Arnon. Here again, one can imagine that changes in the physical world occurred. Any attempt to uniformly extrapolate back to these dates with no singularities would contain errors. Possibly the greatest miracle of all was the splitting of *Yam Suf*. According to chasidic philosophy the very essence or nature of water was changed for this event.[11]

There were many other events which represented a change in the way natural phenomena occurred, but the above examples should be sufficient to illustrate the point that, apart from scientific discoveries of some of the singularities or departures from the assumed uniformity of physical phenomena, the Torah itself contains a number of examples of similar events.

## Acknowledgment

This article is largely an attempt to amplify one of the points made in "A Letter on Science and Judaism," by Rabbi M. M. Schneerson, in *Challenge*, edited by A. Carmel and C. Domb (New York: Feldheim Brothers of Jerusalem and New York, 1976), 142. I have also made use of concepts which I first heard from Rabbis Nissan Mangel and Avraham Pessin.

From *B'Or Ha'Torah* 2, 1982.

# Philosophy

# 17
# God and Rationality

*Yitzchok Block*

## RATIONALITY

The notion of empirical confirmation is no doubt an important element in our concept of rationality. A belief unconnected with what takes place in the world around us and appearing to fly in the face of the reality of the world as we know it cannot be considered a reasonable or rational belief. However, this is by no means the only constraint a belief must satisfy in order to be called "rational."

For example, when Einstein propounded his general theory of relativity in 1916 there was no way the theory could be tested at that time by any experiment. The technical ability to devise such an experiment was lacking. Nonetheless, Einstein was convinced that the theory was true, and when it was finally verified by an observation in 1919 taken during an eclipse of the sun, and Einstein was informed by telegram, he is reported to have said that this was no surprise to him. The theory had such unity and power that he could not imagine it not being true. Was it irrational of Einstein to believe this? Furthermore, was it irrational of him to pursue throughout the latter part of his career a unified field theory that would combine relativity theory with quantum mechanics in spite of the fact that there was no evidence for this theory whatever? The reason he pursued it was that it would bring a unity into diverse aspects of physics where there appeared to be none. He was actually derided by his colleagues for espousing such a theory because there was no evidence for it. It was called a "pipe dream." Einstein's sole motivation was his search for unity. Unity is not the only element that determines the rationality of a belief, but it is certainly a major one and in Einstein's case it was the overriding one.

However, there are situations in which empirical confirmation is primary. For instance, suppose someone has a dream one evening before a horse race that a certain horse is going to win. Most people would say it would be irrational to bet on the horse for that reason. If one wants to bet rationally, he should try to find out more about the horses, their breeding, the jockeys riding the horses, and so on. It might require more research than is worthwhile, but that would be the rational way of going about betting on horses. What if, however, the man bets on the horse and the horse wins; and the next night he dreams about another horse and that horse also wins; and so he continues to dream each night about a different horse which always wins the race the next day. Then after a while it will indeed be the rational thing to bet on the horse he dreams about. This is so in spite of the fact that no one, least of all himself, can explain this strange phenomenon. This shows that in some cases empirical verification can be the prime consideration when determining what is rational to believe. This is so particularly when the main consideration is the achievement of a goal like winning a bet on a horse race, and not the explaining of some phenomenon as in science.

Indeed, there are some who say that the primary function of science is simply to predict events and not to explain them. I think such a view is mistaken. I am sure Einstein thought it was. The primary goal of science is explanation, and the ability to predict is a guide that tells us when our explanations are correct or true. A theory might have unity and aesthetic appeal but could still be false. Thus, predicting events or empirical verification helps determine when theories or explanations are true and thus when it is rational to believe them.

Ability to explain and ability to predict are both necessary conditions that a scientific theory must fulfill in order to be an ideal object of rational belief. Neither by itself, however, is sufficient. In any event, the concept of explanation is separate from that of prediction. One might have two different or mutually contradictory theories that explain an event. What will determine which one of these it is rational to accept will be its predictive ability. However, each of these theories in its own right might have equal explanatory power in different situations, as in the case of the wave and particle theories of light. If we are talking about scientific theories, the rational thing to do is to find the one that has the greater predictive power. However, if one is talking not about scientific theories but about explanations for which empirical verification is not applicable, then what will count as a good explanation will depend upon something other than ability to predict. Surely this is the situation in regard to metaphysical or religious beliefs which cannot be confirmed or refuted by empirical observations, as was argued in my previous article on God and rationality in *B'Or Ha'Torah* 2 (pp. 57–74). However, even in some cases where verification is relevant, the overriding consideration is not always verification.

For example, take a typical case of a Sherlock Holmes mystery. Holmes is able to identify the murderer or the guilty party because he has a remarkable ability to put things together that would escape the notice of most people. The telltale clues are woven together to form a consistent story. It all "hangs together." In other words, a unity is found to hold together a lot of facts that appear disconnected. This explains how Holmes is able to solve the crime. The criminal, of course, confesses, but Holmes knew for certain who was guilty simply on the basis of the power or unity of the explanation he had found connecting all the clues together. At least this is how Sherlock Holmes stories go and we find them pretty convincing.

Even in science, unity or explanatory power sometimes counts as a sufficient condition for rational acceptance. When Einstein was informed in 1919 of the observations made during the eclipse of the sun, confirming his general theory of relativity, he said nonchalantly that he didn't expect anything else. He knew his theory had to be true independently of any observations solely on the basis of its scope, power, beauty, and, above all, the unity that it brought to bear on the physical phenomena of his day. Of course empirical confirmation through successful prediction is necessary, but this is not why Einstein was convinced that his theory was true.

What these examples from the Sherlock Holmes stories and the life of Einstein have in common is that while empirical confirmation is necessary to establish truth, it played little or no role in convincing either of them that they had hold of the truth and that it was not only rational for them to believe that they had the truth but that they were absolutely certain that they were right. They might have been proven wrong, but they thought they were entitled to believe they were right; and the reason for this had nothing to do with verification but with the nature of the explanation itself. That is, it brought all the relevant facts together in a way that gave them unity. In the philosophy of science they call this element of explanation "simplicity," that is, the best explanation or the one it is most rational to believe is the "simplest."

This does not mean it is the easiest to understand or the most obvious explanation. Quite the contrary, often the simplest explanation is the most difficult to discover and requires a sophistication of understanding that goes beyond the superficial. Thus Sherlock Holmes's solution to the crime is not obvious to the police, Dr. Watson, or anybody else who knew about the case. Holmes's revelation of the guilty party comes as a surprise to everyone, but once Holmes explains the reasoning that led him to his conclusion, it becomes obvious, or "elementary," as he often tells Watson, *who* was really guilty. What remains is the man's confession or attempt to escape—but this is incidental to the story. The story is virtually over when Holmes presents the reasoning whereby he was led to the true criminal. It is obvious and brings all the facts together in a way that leaves no doubt about the matter. The reason Holmes's explanation is so convincing is that it is the simplest possible one. That is, it is the only explanation that combines the relevant facts into a single unitary explanation. It has power, for it can explain more facts in the case than any other theory, and it has beauty because it molds a number of diverse elements into a single explanatory scheme.

The more diverse and apparently incompatible the elements, the greater the beauty of an explanation that can, in some surprising way, mold these elements together to form a whole. Beauty in the arts is very close to this notion of beauty in a scientific or mathematical theory, for an important element in this beauty is the unity of a work of art. The more diverse elements the work can weld together in some surprising way, the greater is its beauty. The notions of unity and beauty thus described are the main ingredients that compose what is called "simplicity" in theory construction.

So far as I know there is no argument that can demonstrate or prove that it is more rational to accept the simplest explanation for a number of diverse phenomena, all other things being equal, but it certainly has a strong intuitive appeal. It may be that part of what we mean by rational explanation is that it is the simplest in the sense that it displays the greatest unity and beauty as we have described this.

There is no doubt that this was behind Einstein's conviction that his general theory of relativity was true before he had any empirical confirmation. If it weren't true, then God simply must have made a mistake in forming the world. This is not arrogance but the conviction of a man who has seen the connection or unity of physical phenomena more deeply and more profoundly than anyone had ever seen before. If Newtonian mechanics were the last word in physics, then God would have missed an opportunity of making the world more profound, sublime, and beautiful than He did.

The seeing of unity in the midst of complexity is the key to understanding generally. This is the dawning of the "light," the feeling of "eureka," the flood of understanding that comes when we are wrestling with the solution to some problem. We suddenly see a connection between things. An idea brings them together. We speak in the singular here. We say, "Now I see *it*. I have the *solution*. I understand *it.*" We don't say, "Now I see *them*. I have the solutions. I understand *them.*" This is not just a figure of speech. It expresses the idea that when one understands something, it is a unity—a connection—that one understands, and though the elements connected be multifarious and diverse, the idea that connects them is single. One gets the *idea* for a novel, a painting, a play, a song, or a melody. In all these cases it is unity that makes it what it is. A book can contain thousands of words, many plots, subplots, and characters, but it must have a unity. All these elements must fit together to form a *single* whole or harmony. Otherwise the novel is a bad one or not a novel at all, but disconnected events. The same applies in all instances of art and communication. It is the unity of an idea, a story, a symphony that gives it its power and beauty.

In science and conceptual understanding generally, this unity plays the crucial and central role in drawing the mind to assent. This "drawing" or attraction of mind we call "rationality." The fact that the mind is drawn towards unity and beauty as we have described it simply expresses the idea that rationality and unity of understanding are internally related. That is, the pursuit of unity is part of the essence of what we mean by rationality.

That this is so is indicated by the fact that not everything that attracts assent is called rational. Our minds can be moved to accept as true explanations or theories that are popular or that somehow cater to our needs. One might want to believe the truth of a theory because one has been awarded a large grant to do research on it. Not every inclination of the mind to accept some explanation or theory as true is rational. We mean to restrict rationality to this assent of the mind as it pursues understanding for its own sake apart from "ulterior" considerations. I am suggesting that this pursuit of understanding is nothing else but the pursuit of unity in the midst of diversity, and this is, as a matter of fact, part of the meaning of rationality. It is not the whole of it; not every explanation that displays unity or beauty is true. However, it is certainly rational to believe that of competing theories, the simplest will prove true.

## THE ARGUMENT FROM UNITY

God is the simplest explanation of the world. This does not mean that one understands how God made the world. One would have to be God to understand this. It means that of the various possible explanations of the world or of why anything rather than nothing exists, God is the simplest. Keep in mind that "simplest" here does not mean easiest to understand or something superficial. On the contrary, as we have said earlier, the simplest explanation is one that has the greatest unity and that weaves the most diverse and apparently incompatible elements into a single harmony.

What are the alternatives to God? There are only two. Either the world had no beginning and is eternal, or the world came from nothing—that is, it just popped into existence from absolute nothingness. One cannot say that the world came from some force or power that is not God, for one can ask what is the explanation of that force or power itself. Thus, if one is prone to adopt the Big Bang theory of the origin of the world, one cannot point to some possible previous world or force as the origin of the Big Bang, for one can ask for an explanation of that world or force and so on ad infinitum. The only possible explanation which could satisfy the question, "What is the origin of the world?" is a cause that, by definition, is itself uncaused, in other words, a Being Who is eternal—which is what we mean by God.

Let us rule out the possibility of the world just suddenly popping into existence for no reason whatsoever because this flies in the face of the principle of sufficient reason that everything has some cause or explanation of why it exists. If the world did come from nothing, then the origin of the world is irrational; it defies the principle of sufficient reason. However, why should one admit this possibility as long as more rational alternatives are available? The alternatives are that the world has no origin and is itself eternal, or that God is the explanation of the world. Which of these two alternatives is more rational to believe?

The first question to consider is why anything should exist rather than nothing. The existence of nothing requires no explanation. An explanation is required only for things that exist. The fact that there is a world, therefore, calls for an explanation of how it came about or what brought it into existence.

Some philosophers have argued that no explanation is required for the world in its totality. The principle of sufficient reason might apply to particular events or objects in the world but not to the world as a whole. The reason they say this is similar to the reason why we are not inclined to ask for the explanation of some objects or things in the world. For instance, if you are walking in a forest and see an arrow carved into a tree, you will wonder why this arrow is here, but you will not wonder why the tree is there. It is perfectly normal to find trees in forests, but it is not normal to find arrows carved into them. This is not to deny that there is *some* explanation for this tree being here, but it is not an explanation that anyone would seek out or ask for. It is not something to wonder about, whereas the arrow is. Thus, explanations are required for deviations from a normal pattern, but the normal pattern itself calls for no explanation. In fact there must be some background that is taken for granted before you can wonder about anything. Trees have to be taken for granted before you can wonder about the arrows carved into them.

In a different context, a forester or geologist might wonder why this forest is here or even why this tree is here, but he will have to take other things, such as the germination factor of seeds, soil composition, and climatic conditions, for granted in his explanation. In a wider context, a biologist or weatherman may be interested in explaining the germination of seeds or the weather, but these explanations require other things we take for granted, such as molecular structures, the rotation of the Earth, the angle of the tilt of the Earth's axis from the sun, and the distance of the Earth from the sun. It is impossible to explain literally everything, for then there would be nothing fixed or taken for granted in terms of which everything could be explained. This "everything" is the world (universe) in its entirety. Therefore, it is impossible to give an explanation for the world as a whole, and thus the principle of sufficient reason does not apply to the world. This being the case, it is not possible to ask for an explanation of the world, and one can assume without any qualms that the world is eternal. One then does not need God in order to explain the world, for the world requires no explanation. The existence of the world is what one must assume before one can explain anything in the world.

I think there is a good deal of truth in this. Surely the kind of explanation we are asking for when we wonder why this tree is here or why an arrow is carved in the tree is very different from the kind of explanation we seek when we ask why there is a world at all. To the latter question we don't expect an answer in terms of laws of nature, seeds, and climate, nor in terms of desires or wants that any person could understand, for who can understand the mind of God? "My thoughts are not your thoughts, nor your ways My ways, says God" (Isaiah 55:8).

Perhaps there is a message relevant to man that one can find in the Torah. God wanted a place to dwell in the world below, or God created man in order that man could come to know God, but why does God need a place in this world and why does God need man to know Him? Indeed, why should God require anything other than Himself since He above all else is absolutely self-sufficient and requires nothing?

When one talks about God wanting or needing anything, this is to be taken with a grain of salt. We speak "analogically" for, as we say, who can understand God? When the biologist or soil scientist explains how seeds germinate or how climate affects the growth of plants and trees, we say, "Now I understand." There are, nonetheless, elements in the explanation that are not really understood.

There are no perfect explanations that explain everything because of the reasons mentioned above. Nonetheless, there is a lot to learn about biology that could take a lifetime to master. However, when one asks why there is a world and in reply is told that God created the world, does one then understand *how* the world was created? Could one study this for any length of time? After one has said that God made the world is there anything left to add which will enable one to understand it better? It seems not. For these reasons there is something to be said for the arguments of those philosophers who claim that the world neither has nor requires an explanation for its existence, in the sense that one can understand how it came about.

In spite of all this, there is something important that the philosophers have left out of their analysis. This is the simple fact that people *do* wonder where the world came from. The human mind *is* capable of asking such transcendental or ultimate questions, even though there may be no clear answers of the kind we expect from such questions as where this tree came from. The philosopher says such questions have no answers, so it is better not to ask them; but the minds and hearts of men do not always obey the decrees of philosophers. Furthermore, this is not some technical or abstruse question which one could ask only after having studied philosophy for a few years, as are many questions asked in philosophy. It is a question asked by simple folk as well as the most learned, and it is a question that is important to them. It could affect how they live their lives and their whole attitude towards life and the world. They could become different people according to how they answer this question. The existential implications are considerable. This is not the case with many questions in philosophy today which have as much relevance to human life as the proverbial, irrelevant medieval question of how many angels can dance on the head of a pin. The question of how the world came about and why is a real question that almost every human being asks, and it will not go away because some philosophers call it "transcendental" and turn thumbs down on it. In other words, though the question may have no answer that we can understand, this will not prevent us from asking it.

It is part of the human condition to wonder about this as well as other transcendental questions. We are arguing here that once one poses this question, the most rational answer is God. The atheist will choose not to ask the question in the first place. This is what he is doing when he says the world itself is eternal, and he will support this with arguments of the kind spelled out above. However, none of these arguments removes wonder about the origin of the world from the minds of the vast majority of people. The atheist must consider himself one of those select few who has succeeded in ridding himself of the question and thus of any need to supply an answer.

Insofar as one is troubled by this question at all, God is the only reasonable answer. As we have said, this does not mean that one understands how God made the world. It simply means that He is the origin of the world, and therefore certain elemental requirements of human understanding are fulfilled by this in spite of the fact that we do not understand how He made it.

It is at this point that what I call "the argument from unity" becomes relevant. The basic thrust of rationality is, as we have said, the pursuit of unity. This is true of every human endeavor, whether it be in science, art, music, or even politics. It is also true on a transcendental level in terms of understanding the world as a whole.

Unity of this level is expressed by saying that the world is perceived as one world—not two, three, or even an infinite number. Another way of expressing this is that everything in the world is connected in some fashion with everything else. It may be difficult to fathom the idea I am trying to get at here, for the concept is so intuitively a part of our thinking that it is difficult to imagine a world which is not "one." Philosophers who are good at imagining improbable situations have devised situations in which the world in some important sense might not turn out to be "one." Here are a couple of examples.

1. David Hume suggested that there is no proof that the laws of nature might not radically change at any moment so that all your past experience would be no guide whatsoever for the future. There would be no reason for the change or explanation as to why or how it occurred. It is simply that the way in which things happen would radically change and that would be the end of it. In that case the world would be two, not one. The history of the world up to the time of the change would be one world, and the history of the world from that time on would be an entirely different world. There may be no survivors from one world to the other, but if there were, they would have to "retrain"—if retraining would make any sense. Suppose, for instance, the constants of nature were to change or reverse in some odd way.

Suppose gravity became a force of repulsion instead of attraction, or the speed of light slowed down to 5 miles per hour, or that magnetic forces ceased to function altogether. Could such a world be in any way continuous with the world we know? It would be a different world altogether. If there were such worlds so radically different from ours somewhere in outer space, we could say right now that the universe as a whole is two and not one.

Hume argued convincingly that the rationality of man has no proof that this hypothetically bizarre turn of events can't or won't happen at any given moment. This is correct, but he drew the erroneous conclusion that man's rationality is not justified in making any prediction about the future. According to Hume, you have no rational ground for choosing to eat an egg for breakfast instead of the fork. In fact the situation you are in from a rational point of view when you sit down to breakfast is that you have no idea *what* to eat. The fact that an egg always satisfied your hunger in the past is no evidence whatsoever that it will do so today or any time in the future. Under such circumstances—when you cannot make up your mind whether to eat the egg or something else, say, your fork—the "rational" thing to do would be to flip a coin.

This is patently absurd. It is, of course, rational to eat the egg and not the fork because it is rational to suppose that the world will not radically change at any moment. The reason it is rational to suppose this is that it is the nature of the rational mind to conceive of the world as one and continuous. The world may indeed prove discontinuous, but it would then become incomprehensible. Our rationality would be unable to cope with it. It would become a "mad" world.

2. Descartes argued that one cannot prove that one is not dreaming even when one thinks one is awake, and what we call "waking life" might be one long dream. This is true. One cannot prove that life is not a dream. We may in fact some day "wake up" from a waking life to discover that it is all a dream. Nonetheless, it is not rational for a person to think he may be dreaming when he knows he is wide awake. The reason why it is rational to think that waking life is not a dream is that if it were, then we would be living all the time in two worlds, not

one. Namely, there is the world of waking life in which we are born, grow up, are educated, make plans, have friends, raise families, achieve goals, write books, acquire wealth and possessions, and finally die, leaving our descendants behind to inherit that wealth. If all this is a dream, we do none of this. We are not born, do not grow up, become educated, or acquire wealth, and when we die we cannot leave wealth to anyone. However, we certainly do think about our lives in this way. We think it important to achieve goals, raise families, acquire wealth, and so on. Even if we thought we were dreaming all the time, we would have to live our lives as if they were "real." It would do no good to argue with our families or our stomachs that their or our hunger pangs were nothing to be concerned about since it is all a dream. We would still have to go to work to put food on the table.

Therefore, if it is true that we are dreaming, then there are two worlds that are unconnected, so far as we know. There is the world we live in which is a "real" world and there is the "dream" world which Descartes thought we might be living in, and these two worlds do not impinge on one another at all. They are completely different worlds and you cannot infer from one to the other as we ordinarily do in relating our real dreams to waking experience. We can of course relate our real dreams to our waking experience. Psychologists and psychiatrists do it and we recognize it ourselves when we dream about things we have desired or feared in our waking experience. As we ordinarily know it, our dream life has an intimate relationship with our conscious and unconscious mind. They are not two different worlds that have no effect on one another. They are various aspects of one life and can be understood in terms of one another.

The kind of situation that Descartes asks us to ponder is a different one altogether. The world that we think is the real world might in fact be a dream, but if it were, there would be no way of discovering this. We would simply have to live our lives as if they were real. And if they were dreams, so be it. The one thing, however, that would not be possible in this situation would be that our dream world could have implications for our waking life or vice versa since by definition they are two worlds that impinge on one another at no point.

This is similar to the two worlds envisioned in Hume's suggestion that the world we know might suddenly change in a completely
surprising and unsuspected manner such that there could be no way
of understanding how one came from the other. No theory could explain it.

In both these situations we say that it is rational not to expect
the world to change unexpectedly and it is rational to think one is not
dreaming all the time. The reason in both cases is that it is rational to
think of the world as one in the sense that we have expressed it. One
could stop here and be content, if one wished, with a statement about
the psychological makeup of the human intellect. It is such that it just
takes unity as its rational goal. However, one could go on to argue that
a ground of this unity is God in that He is one and His unity with the
world is one. Since He is the Creator of human reason also, this explains
why the mind tends to seek unity. The search for unity is none other
than the search for God Who is the ultimate unity of all things. This
does not prove the existence of God. What it proves is that it is rational
for the mind—a mind which by its nature seeks unity.

The role that God plays in the pursuit of understanding the world
can be brought out by imagining a Godless world. Suppose God had
not created the world, God forbid. Then it would not be rational to
expect that the laws governing the world could ultimately be unified
under a single theory or schema or explanation. If the world exists by
accident, which is the alternative to God, it would not be reasonable
to expect the world to be unified in its vast variety of animal, plant,
and inorganic substances any more than it would be reasonable to
expect marbles rolled indiscriminately on the floor to form some recognizable pattern. Things that happen by chance or by accident do
not show regularities that indicate the presence of laws of nature. That
there are regularities in nature indicates the existence of laws, and we
expect that the laws of nature work together to form a single universe,
not two, three, or more separate worlds that have no relevance to one
another or that might succeed one another in the manner suggested
by Hume. There is ultimately only one world. All science and basic

rationality of man take this for granted, but there would not be the slightest reason to believe this if the world were an accident.

This, I believe, is also what Aristotle meant when he said that the fact that nature displays regularity shows that the world is not an accident (*Physics* 196a 29ff). Hume, on the other hand, begins with the assumption that the laws of nature are accidental. If one begins with this assumption, there is no good reason for not supposing that the world might not suddenly change by accident again. "Easy come, easy go," as the saying goes. Arguing backwards, one could say that since we do think it rational not to expect such an accident, it is rational to suppose that the world did not come about by accident and that its dependency on God is the precise supposition that is required for a ground of this rationality, in case you like to have your rationality "grounded." This is a rational position, for as human beings we want to think of the world as a whole, a unity, that is, as one. The various forces and laws that hold the world together must be working in harmony with one another. We find this idea almost irresistible. It is the simplest, most rational way of looking at the world.

It is not necessary that the world be one. It may well be that two or three different incompatible systems compose what we call the world, and there is really no *one* world. However, until this is shown to be true, I doubt if anyone would entertain the idea seriously. Why does the idea of a unified world appeal to us? Why have philosophers always tried to give a single explanation for all phenomena? From the dawn of scientific speculation among the ancient pre-Socratic philosophers who said the world was essentially water, fire, or air, or some other element, down to Einstein's speculations about a unified field theory, there has been a continuous attempt to give a single thematic explanation of the universe. The wonder of this attempt is that it has not been thwarted by nature herself. She has always yielded more and subtler forms of unity to those who have searched for it and has even rewarded those in pursuit of a unified field theory.

Whether or not this pursuit will prove fruitful is not relevant to the fact that it is the rational approach to nature, for it is the nature of reason to pursue unity. Should this search not succeed, we shall, with a sigh, have to admit that nature is not as rationally laid out as we would have liked. There is no necessity that nature obey all the intricate mathematical calculations that physicists make. Physicists do expect that nature will obey the laws of mathematics. This is nothing more or less than the expectation that nature does follow a rational plan and that, as such, nature will find the simplest way of doing what has to be done. Is not the simplest way none other than the way that brings all the phenomena of nature under a single system that forms a single harmony, a oneness, and a unity? Has this not been the pursuit of man since the time he began speculating on how to understand the secrets of nature? If one grants this, it is only a small step to attribute this unity to God Who is the ultimate unity of all things.

The central clarion call of the Jew is the oneness and unity, rather than the mere existence, of God. The Jew does not say, "Hear O Israel, the Lord our God, the Lord exists," but rather, "Hear O Israel, the Lord our God, the Lord is *one.*" What the Jew proclaims is the oneness and unity of God, not only with Himself but with the world. The unity that the scientist finds in the world is but a small glimpse into the unfathomable oneness and unity of God with the world. Mystics have a glimpse of this unity when in a flash of insight they see the unity of all things— that everything really is one.

The external form of this unity is expressed in the laws of nature, but the essential form of this unity is felt in the fusion of all things with God. The world is not one thing and God another, but the world and God are truly one, as Abraham proclaimed when "he called the name of God, God of world" (Genesis 21:33). Not God of *the* world as if God were separate from the world, but God of world where God and world are inseparable.

Perhaps this is the intention of the psalm that we pondered previously, "the heavens proclaim the glory of God and the firmament tells of the work of His hands" (Psalms 19:2). The unity that one finds in the heavens is a small reflection of that ultimate unity of God and the world which is the ultimate truth.

This truth may not be logically necessary, but it is more rational than assuming that there is no single world or that all the indications of some astounding unity in the world are nothing more than misleading signs. That the world is one by accident is too improbable for any rational person to believe. There is no avoiding it. It is rational to think of the world as one, and therefore it is rational to accept God as the ultimate Being or ground for the world.

From *B'Or Ha'Torah* 6E, 1987.

# 18

# Musings on
# (the Logic of) Repentance

*George N. Schlesinger*

## I

The Talmud (*Yoma* 86b) attributes to Resh Laqish the view that when a sinner repents—not out of fear but out of a loving desire to do the will of his Creator—he is granted full pardon. And not only is he forgiven, but all his previous transgressions are accounted as virtues.

In *Kiddushin* 40b, this strikingly generous rule is said to apply even to a person who has led a life of practically unmitigated vice; as long as before drawing his last breath he thoroughly regretted his sinful ways, he is granted full pardon and assigned a high spiritual rank.

The question likely to arise in many people's minds is: How are we to reconcile this pivotal idea with our belief in Divine fairness? On the surface of it, Resh Laqish's doctrine appears to violate our natural sense of justice which demands that the gain be proportional to the pain. One who regrets his wrongdoings, however sincerely, at a time when he is conscious of his imminent death can scarcely be regarded as making a substantial sacrifice or as having undergone a protracted, arduous process of repentance. He knows that he will have no further opportunities of succumbing to the blandishments of the Evil Urge and is therefore safely beyond its reach. Moreover, by this time he has probably lost those cravings that led him astray in the first place. Thus we may wonder: Is justice well served if Resh Laqish's immense reward can be acquired with so little trouble? Is, among other things, Resh Laqish fair to those who had to earn their way, through a lifelong strenuous struggle, by sustained hard work and self-denial?

## II

Many passages from the Talmud Sages support the view that nothing of value can be acquired without toil and trouble. For example, in *Berakhot* 5a Rabbi Shimon ben Yohai is quoted as saying, "Three good gifts did the Holy One be blessed give to Israel, and all were given only by means of *yissurin*, and they are: Torah, the Land of Israel, the world to come."

The key word is *yissurin* which has been variously translated as "torment," "affliction," "agony," "pangs," "pains," and "punishment." However, Reuben Alcalay, in his highly acclaimed dictionary, renders the phrase "The Torah cannot be acquired save through *yissurin*" as "There is no royal road to learning." It seems reasonable to read Rabbi Shimon ben Yohai as saying not that learning entails excruciating misery, but that every bit of precious knowledge must be obtained through mental struggle. Similarly we may assume that, according to Rabbi Shimon, one's portion in the hereafter must also be acquired through arduous effort. And if there are some "who acquire a place in the world to come in a single moment" it seems reasonable (however little it is given to us to grasp otherworldly matters) that portions differ in size or quality, and that justice would demand that a fuller portion ought be harder to come by.

There are several possible ways of eliminating this apparent gross violation of our natural sense of justice whereby a small effort and one a thousand times as great obtain the same reward. One of the more convincing approaches assumes that though the act of contrition may last only a few moments, the psychological strain and effort it involves is nothing less than heroic. A last-minute repentance, to be genuine, must be accompanied by the sinner's full acknowledgment of having led a hollow, weary, stale, and unprofitable life, of having spent all the precious time allotted to him in the pursuit of unwholesome, fleeting pleasures. Thus the authentic *baal teshuvah* (penitent) is much more than, as Coleridge would have it, "a sadder and a wiser man;" he is a self-convicted, self-debasing man with a devastated self-image, whose soul is cleansed and ennobled by the pain of his crushed ego. In view of this profound pain, we should perhaps no longer wish to question the fairness of the "death-bed" penitent's reward.

The issue is not so simple as all that, however. In fact, it is much more complicated than many of us would like it to be. Fortunately, perhaps, while the act of *teshuvah* (repentance) is certainly mandatory, a full understanding of the logical intricacies surrounding it is most probably not. Be that as it may, here I should like to make a small contribution toward the elucidation of this difficult problem.

I propose to begin with an argument which seems to show that the concept of "eleventh hour" repentance implies what is known as an infinite, vicious, circular regress. However, first let me attempt to provide an elementary outline of this fundamental notion. Though as far as I know no one has applied it to our problem, infinite regresses in general, both circular and noncircular, have deservedly received considerable attention from logicians past and present.

# III

An ancient and probably the most widely known instance of a vicious, infinite, circular regress is associated with the name of the Greek philosopher Eubulides of the School of Megara, who is said to have asked whether a man who declares: "I am now lying" ought to be believed. Clearly if we believe that he is now lying, then we must treat his statement "I am now lying" as a false statement. But then, if that statement is false, he is not lying, in which case, of course, he is telling the truth and is lying, and thus he is not . . . , and so on ad infinitum. Over the centuries logicians have suggested innumerable ways to break out of this vicious circle; none of their solutions will concern us here.

Eubulides's liar's paradox is an important landmark in the history of ideas, even though many highly intelligent persons have felt no passionate desire to investigate it and have even tended to dismiss it as an overrefined piece of casuistry. Nevertheless, the attempt to solve it has stimulated inquiry into such important topics as self-reference, meta-statements, the theory of types, and so on.

A much less known example of a circular regress occurs in the story of a king who, on January 1, 1234, decrees that a serf be released immediately from the bondage of the nobleman to whom he belongs, on the condition that in the course of the coming year the serf transgresses no law of the realm, voluntarily or involuntarily. According to the king's decree, the status of the serf during the whole year of 1234 remains indeterminate. But if at midnight on January 1, 1235, it turns out that he was completely law-abiding throughout the preceding year, then it will become evident that he has been a free person throughout the entire year. Otherwise, he has been and continues to be a serf. Now it happens that on June 1, 1234, the serf sells his horse to a neighbor. His master, the nobleman, on hearing about the property transfer, cancels the sale, using the privilege granted to him by law to nullify, if he so wishes, any of his vassals' legal transactions. On the next day, the serf works with the horse without asking permission of the man to whom he had sold the beast the day before. Apart from this one problematic act, the serf has been law-abiding throughout the whole year.

Is he to be declared on January 1, 1235, a freeman who has been released from serfdom since the beginning of 1234? If we say that he has been free the entire preceding year, then of course the nobleman had no right to cancel the sale of his former serf's horse. Thus, on June 2 the seller was working without permission with a horse belonging to his neighbor, thereby transgressing a law. Consequently, he is not free. But if he is not free, then the nobleman did have the right to invalidate the sale and, therefore, on June 2 the serf (?) was working his own horse, not transgressing any law and thus becoming free on the coming January 1, in which case he is not free . . . and so on.[1]

Former Chief Rabbi Amiel of Tel Aviv, the leading exponent of the uniquely rigorous logical method of Talmud study devised by the great *gaon* (genius), Rabbi Shimon Shkopp, once constructed a halakhic (Jewish legal) example that has precisely the same structure as the circular regress of our king–nobleman–serf story. Suppose Dan hands a bill of divorcement on the new moon of the month of *Shevat* to his wife Dina—who is anxious to have their marriage dissolved— saying: "You cease to be my wife as of today on the condition that during this entire month you do not violate any law, even unintentionally." Dan explains that Dina's marital status during the month of *Shevat* remains unknown; however, if at the very end of *Shevat* it is confirmed that she has successfully avoided all transgressions in the last thirty days, then it becomes evident that she has been a merry divorcée throughout the month (such arrangements are not allowed, but that does not affect the logic of the situation). Complications begin when on Shevat 5, Dina makes a *neder* (vow) to eat nothing but raw vegetables for the next twenty-five days. She does it for added protection against the invalidation of the bill of divorcement, in view of the fact that the *kashrut* of almost any food can be called into question. Dan learns about the *neder* on the next day, at which time he uses his presumed husbandly prerogative, declaring Dina's *neder* null and void. Subsequently on *Shevat* 15 (the New Year of the Trees), Dina indulges in a few oranges and apples. Otherwise she scrupulously observes all the *mitzvot* (commandments).

At the commencement of the new month of *Adar*, are we to regard their marriage as having been dissolved thirty days earlier? Suppose the answer is yes. In that case, on Shevat 6 she was no longer married and her former husband Dan had no power over her vows. Thus she is bound by the *neder* to refrain from eating anything but raw vegetables. Thus, celebrating Tu Bishvat with apples and oranges amounted to violating the law requiring vows to be kept. It follows that she has not fulfilled the condition attached to the bill of divorce-

ment and remains married to Dan, who did after all have the authority to cancel her vow. In that case, however, Dina ate nothing prohibited and she became a divorcée from the moment she received the
bill of divorcement on the first of *Shevat . . .* and so on.[2]

It may not be so surprising to find that some concepts in the
halakhic domain can be treated as problems in abstract logic. But it is
different with *makhshavah* (thought). Some might view the necessity
or even the mere possibility of applying arid, rigorous logic within the
sphere of beliefs and opinions, where faith and emotional commitment
should play the predominant role, with a certain amount of misgiving.

It is undeniable that eloquent rhetoric, employing poignant para-
bles, poetic images, and suggestive allusions, is far more likely to be an
effective means of inspiration and of persuasion than elaborate, dis-
passionate, analytic reasoning. However, an honest attempt to argue
patiently in conformity with the exacting standards of logic is what is
most likely to lead to genuine clarity and insight. Many will agree that
understanding just for the sake of understanding is a precious enough
aim to be pursued.

With this in mind, let us look once again at what has been said
to constitute a higher form of repentance. Recall that we suggested
that the reason why even a thoroughly wicked individual who turns
around only in the last moment, and does so out of love, rises to the
level of a fully virtuous person is that he has undergone the excruciating experience of unflinchingly facing the fact that his entire life, now
about to end, has been a sham. He is admitting to himself the irre-
trievable loss of the one and only one chance mortals are given to self-
enhancement; he has come to the agonizing realization that he has
recklessly squandered the precious earthly years allotted to him.

Now, however, we are in a position to ask not whether the principle governing repentance is or is not fair, but whether it is altogether
logically coherent. Let us begin with the assumption that all the
penitent's sins retroactively turn, as Resh Laqish maintains they do,
into virtues. Let us also assume that our *baal teshuvah* has a rudimen-
tary knowledge of the basic principles of repentance and is thus aware
that his past life has been instantaneously transformed and is now full

of virtuous acts. It immediately follows that his repentance does not after all involve any acknowledgment of having led a life devoid of meaning, since now he realizes that his past life is replete with righteous works, he is not subject to the painful awakening to dreadful loss. Therefore, he cannot be considered a genuine *baal teshuvah,* and assuming he is a reasonably intelligent person he is bound to grasp this bitter fact as well and arrive at the inevitable, distressing conclusion that he is to die as a wretched sinner. To be aware of such a conclusion is, however, as we said earlier, sufficiently spirit-crushing to transform his life into a highly virtuous one, which in turn implies that he has nothing to be overly distressed about and has therefore not really earned the glorious reward promised by Resh Laqish . . . and so on ad infinitum.

## IV

I do not wish to imply that I have generated an insoluble paradox; however, I do believe that the issue warrants serious thought, and the problem does not yield to casual treatment. An example of a somewhat facile solution would be to distinguish between penitents with different degrees of knowledge. The paradox arises only in the case of a perfectly informed sinner, who at the very moment of experiencing the deep regret about the vacuity of his life is fully aware that his remorse removes the grounds for regret. Perhaps Resh Laqish is talking about someone ignorant of, or merely vaguely acquainted with, the workings of *teshuvah.* For such a person no difficulties would arise.

Among the inadequacies of such a suggestion seems to be its implied penalization of the knowledgeable person. As a rule, ignorance in the spiritual domain is regarded not a bliss but a serious fault, and thus the learned ought to be better and not worse off than the unlearned.

It has also occurred to some to argue that the paradox seems based on the questionable assumption that one is unable simultaneously to experience joy and sorrow, both originating from one and the same source. An ascetic, for example, who abstains from food and drink, may endure the pain of hunger and thirst, yet at the same time derive considerable satisfaction from his transcendence in self-mortification. Thus there may be no grounds for assuming that a sinful past could not possibly cause deep regret for one reason and, at the same time, intense joy for another.

This reply does not work, as it overlooks a subtle difference between the two cases. The ascetic's pain comes from his experience of hunger and thirst; his pleasure comes from something else, namely, from his awareness of an ability to endure that pain. Not only are these two mental states compatible, but the former is a necessary precondition for the latter. But a person's past life is either full of sin or it is not; it is logically impossible that it be filled with sin and yet also free of all sin. But in the former case there is absolutely nothing to be joyous about, and in the latter case there are no grounds for regret and shame and thus no grounds for forgiveness.

## V

Let me now embark on an outline of an approach that seems to me more promising. It is based on two presuppositions. The first is the general assumption that an intricate and difficult task may occasionally be accomplished swiftly and easily—if the performance has been preceded by long and arduous preparation through which the ability to do the difficult at once has been acquired. The second assumption, more specifically relevant to the matter at hand, is that *teshuvah me'ahavah* (repentance from love) accomplished in a few brief moments is, as a rule, a manifest culmination of extended, strenuous, and unexposed spadework. I shall try to explain.

Deuteronomy 10:12 says: "And now Israel, what does the Lord thy God require of thee, but to fear the Lord thy God, to walk in all His ways, and to love Him. . . ."

*Berakhot* 33b raises the question: How could Moses imply that the fear of God—which, as has often been pointed out before, is a much higher form of piety than merely being afraid of Him—is so easily accomplished as all that? To which the Talmud replies: To Moses himself it was indeed an easy matter.

Prior to a closer look, one might be wondering: Was Moses our Teacher so self-centered as to be unable to appreciate how hard most of us find the assignment "to walk in His ways"? Moreover, are we to assume that Moses reached the heights of piety without any serious effort? Surely, if that were the case, he could not deserve the immense veneration in which he is universally held!

We have all heard of the engineer who was asked to repair a complicated computer system. When asked how he could demand $1,000 for his services since all he did was push a single button, he explained that indeed he was asking no more than a single dollar for pushing the button, but that he felt entitled to charge $999 for knowing which button to push! Many years of industrious study were invested by the engineer who now knows his way around an intricate piece of machinery, just as a famous concert pianist's precise and fluent rendition of a complex musical passage had to be preceded by innumerable exhausting rehearsals. Similarly, a great *tzaddik* (righteous person) may act now with great alacrity; his performance may call for no effort. Indeed, given his present elevated state, it is bound to be a source of fulfillment and pleasure. However, pain and toil had to be invested, even in the case of a *tzaddik* of the highest order, like Moses, throughout the strenuous process of raising himself to his current level. Moses tells us, however, that such a level is accessible to each one of us.

## VI

On the view just presented, the original paradox could be claimed to have vanished. It could well be said that *teshuvah me'ahavah*—ideal repentance that may justifiably serve as a source of joy at having transformed all past sins into virtues—is hardly attainable in a brief moment. Such a reversal could not be expected to take place without a prior, protracted, arduous, anxiety-filled process. It is only on the completion of the inner struggle that the prolonged process of *teshuvah* may be said to have come to its happy conclusion. Thus when the Talmud Sages imply that *teshuvah* may be attained in an instant, they may have been referring to the culmination of the process, when the feelings of remorse and the determination to change have ripened so as to rise to the surface and fill the penitent's conscious mind entirely. In other words, once the preparatory stages have been completed, the last and fully conscious stage in an individual's regeneration may be very brief. But the early stages are characterized by growing feelings of regret and frustration, since there is no basis for gratification before the conclusion of the process.

We can strengthen our point if we remind ourselves of the basic doctrine cited by Maimonides according to which *teshuvah* consists of three components: (1) acknowledging the sin; (2) regretting it; and (3) resolving never to repeat it. In view of what has been said, we need to assume that these parts cannot be accomplished simultaneously. Regardless of how long or short the duration of each element, as long as they can be completed only serially the original paradox does not arise. There would be a problem if it were indeed the case that sincere regret is both necessary and sufficient to ensure *teshuvah*, and genuine *teshuvah* guarantees that there are no grounds for regret. This surely would generate the absurdity that by eliminating the reason for self-reproach, repentance removes an indispensable condition for its own existence, that the very presence of *teshuvah* ensures its absence! But this argument is based on an error: sincere regret is merely stage 2 in the process of *teshuvah*. It is thus a necessary but not a sufficient condition for *teshuvah*. Thus, at no point throughout the entire period in which an individual accomplishes stage 2 is there any reason for rejoicing, since repentance has not yet been achieved, and thus there is nothing to assure him that the grounds for his regret have disappeared. It is only later, at the final moment of a hard-won triumph when he accomplishes stage 3, that the *baal teshuvah* will have earned Resh Laqish's sublime reward.

From *B'Or Ha'Torah* 7E, 1991.

# Archaeology

# 19
# Ancient Synagogues and the Temple

*Asher Grossberg*

In its *halakhot* (laws) on how a Jew should prepare himself to pray in a synagogue, the *Shulhan Arukh* (the Code of Jewish Law) says: "One enters an extent of two entrances and then prays" (*Shulhan Arukh, Orah Hayyim, Hilkhot Tefillah* 90:20).

The above *halakhah* (law) comes from a statement by Rav Hisda in the Babylonian Talmud (see "Chronological Order of the Sources and Personalities Quoted" at the end of this volume). We shall compare the different versions of Rav Hisda's statement and various interpretations expressed by other Talmud Sages and, later, rabbinic authorities that are presented by the *Shulhan Arukh* in its formulation of this *halakhah*. Then Rav Hisda's statement as rephrased in the Jerusalem Talmud and the *Midrash Rabbah* will be studied here in the light of archeological findings at ancient synagogues in Israel. These synagogues will be compared to the edifice of the Second Temple, and finally the meaning of this *halakhah* will be discussed.

237

## THE SYNAGOGUE ENTRANCE AND ONE OF THE LAWS OF PRAYER

Rav Hisda said: "One must always enter two entrances into the synagogue." (Babylonian Talmud, *Berakhot* 8a)

This opinion evokes surprise in the Talmud: "Two entrances? What do you mean?" Then the question is resolved by rephrasing the statement: "*Rather say, a distance of two entrances and then pray.*"

The above short discussion in the Babylonian Talmud was given two major interpretations: one by Rashi and one by Rabbeinu Yona.

1. Rashi says that the worshipper must enter into the synagogue a distance the width of two doors and not sit by the entrance. He explains that sitting in a synagogue should not seem a burden or bother. The Rambam (Maimonides) in the *Mishneh Torah, Hilkhot Tefillah* 8:2 quotes Rav Hisda and the ensuing discussion in the Talmud found above, and apparently his opinion is that of Rashi. Rabbeinu Yona allows an exception to Rashi's reasoning. He says that when one's permanent seat in the synagogue is located near the doorway, then one is allowed to sit near it. The explanation of Rabbi Meir of Rothenburg (brought forth in the *Tur*) is that sitting next to the door is prohibited because one is liable to look outside rather than concentrate on praying. Thus one is allowed to sit by the door if it does not open onto the street.

2. Rabbeinu Yona in the name of other commentators, as well as Mordekhai and the Rosh, interprets the "two doors" as meaning a duration of time rather than a distance in space. That means that one must wait for a period of time equivalent to going through two doorways before praying.

These two opinions, including the two modifications on the prohibition against sitting by the door in the synagogue, are presented as law in the *Shulhan Arukh*:

> One enters an extent of two entrances and then prays. There are those who interpret "an extent of two entrances" as meaning one must go

*eight tefahim (handbreadths) inside the synagogue, so as not to sit by the doorway and make synagogue attendance seem onerous.* This does not pertain to someone who has a special [permanent] place by the door. And there are those who say that the reason for this is that *one cannot concentrate if one is looking outside,* and so according to this reasoning this *halakhah* does not pertain when the doorway does not open into the public domain. *And there are those who say that one should not rush to pray but should pause* for an extent of time equivalent to entering through two doorways. *And it is correct to heed all of the opinions. (Shulhan Arukh, Orah Hayyim, Hilkhot Tefillah 90:20)*

### The Jerusalem Talmud Version

Rav Hisda's statement in the Babylonian Talmud cited above is presented after commentary on the verse: "Happy is the man who hearkens to Me and is diligent at my doors daily and guards the doorposts of My entrances" (Proverbs 8:34). The commentary in the Babylonian Talmud revolves, therefore, around the plural end of the verse: entrances. In the Jerusalem Talmud, Rav Hisda's statement is changed to emphasize the plural word *doors*, which appears at the beginning of the verse: "Rav Hisda said: One who enters a synagogue *should go inside through two doors.* Why?' 'Happy is the man who hearkens to Me, being diligent at my doors daily.' My doors, not my door; doorposts, not doorpost" (Jerusalem Talmud, *Berakhot* 5:1).

The emphasis of the Jerusalem Talmud on the word *doors* rather than on *entrances* supports Rashi's opinion that the reference is to the distance which one should traverse upon entering a synagogue and not to the time that should pass before one starts to pray. (This is what the Rosh wrote on *Berakhot* 7:1.) The *Tur* presents the Jerusalem version in order to defer the opinion of Rabbi Meir of Rothenburg that the obligation to enter all the way inside pertains only when the synagogue entrance faces the street. In the Jerusalem Talmud no distinction is made between the various cases. As has been shown, the *Shulhan Arukh* presents all the opinions and deems it correct to heed all of them.

The Bach had a different explanation to the Jerusalem Talmud version. He teaches that "*Every synagogue must have a vestibule in front so that it is entered through two doors.*" That is to say, one should not pray in the hallway but rather one must enter inside the synagogue. The Bach supports his argument with the *Midrash Rabbah* version. His explanation is also supported by archeological evidence.

### The *Midrash Rabbah* Version: Archeological Evidence and the *Halakhah*

This is what the verse says, "Happy is the man who hearkens to Me and is diligent at My doors daily. . . ." What is the meaning of "being diligent at My doors"? The Holy One Be Blessed said, "*When you come to the synagogue to pray, do not remain standing at the outside entrance, but see to it that you enter through the inner door*" for the verse does not say "diligent at My door" but "at My doors," i.e., two doors. (*Deuteronomy Rabbah, Ki Tavo* 7:2)

Thus the Midrash version teaches us that the entrance to a synagogue was through two doorways, one inside the other. This can be done where there is an outside vestibule leading into the sanctuary, as the Bach said. Archeological findings confirm that some of the early synagogues indeed were built in this manner.

There are two types of entryways found in the ancient synagogues of Israel: (1) entrance in the wall opposite the direction of prayer or in one of the two side walls and (2) entrance in the wall facing Jerusalem.

In a number of the synagogues of the first group is found a narthex (narrow hallway) in front of the entrance. The narthex was entered through a narrow side—as can be seen in the synagogues of Ein Gedi and Hammat Tiberias (Figures 19–1 and 19–2)—or through an opening opposite the entrance to the sanctuary—as at *Beit Alfa* (Figures 19–3 and 19–4). Synagogues built in this style illustrate the meaning of the passages studied above from *Midrash Rabbah* and the Jerusalem Talmud version.

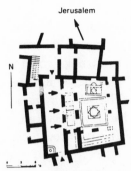

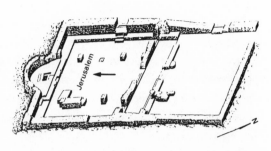

**Figure 19–1.** Ein Gedi, plan of the synagogue (5th–6th centuries). From *Qadmoniot* 5, 1972, p. 52. Used by permission of the Israel Exploration Society.

**Figure 19–3.** Isometric reconstruction of the synagogue at Beit Alpha (6th century). From E. L. Sukenik, *The Ancient Synagogue of Beth Alpha* (Jerusalem: Magnes Press, 1935).

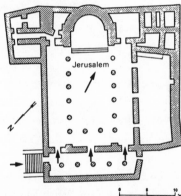

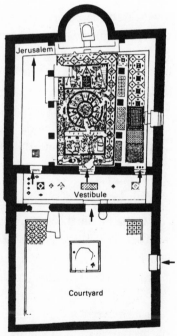

**Figure 19–2.** Plan of the synagogue at Hammat Tiberias (Umayyad Period, second half of the 7th century). From *Qadmoniot* 1, 1969, p. 123. Used by permission of the Israel Exploration Society.

**Figure 19–4.** Plan of the synagogue at Beit Alpha (6th century). From *Qadmoniot* 5, 1972, p. 42. Used by permission of the Israel Exploration Society.

The Bach's conclusion that a hallway should be built in front of the entrance to the synagogue sanctuary was brought up in the seventeenth century *Magen Avraham* as a reason for constructing the synagogues of that period according to a similar plan: *"Thus they were accustomed to make a vestibule in front of the synagogue."* The *Mishnah Brurah* adds: *"And so they used to do everywhere as today, and therefore it is good from the very outset to pay respect and pray not in the courtyard but in the sanctuary"* (*Hilkhot Tefillah* 90:20).

## THE SYNAGOGUE ENTRANCE, THE TEMPLE ENTRANCE, AND THE NICANOR GATE OF THE TEMPLE COURTYARD

The explanation given by the Bach for building a vestibule in front of a synagogue is so that the synagogue will be like the Temple, which had a vestibule leading into its sanctuary. However, statements based on the physical plan of the Temple and the Tabernacle, made by the Talmud Sages on synagogue entrances, relate to the location of the entrance but not to its structure: *"Synagogue entrances open only toward the east, as was found in the Sanctuary* [the Tabernacle] *that opened to the east, as was said* (Numbers 3:38): *'Those who were to camp before the Tabernacle, in front—before the Tent of Meeting, on the east . . ."* (*Tosefta, Megillah* 3:14).

Synagogues with entrances in the eastern wall have been found so far mainly in the south of Mount Hebron, in Samu (Eshtemoa), Khirbet Susiya (see Figure 19–5), and Horvat Rimon (third-century C.E. stratum), and some in the north of the country such as Khirbet Summaka on the Carmel and the Arbel and Sasa in the Gallil.

In other ancient synagogues the group with entrances facing Jerusalem is that which structurally resembles the Temple. Synagogues of this type are found mainly in the Gallil and Golan, the earliest dating to the third century C.E. They are similar to the Temple in the structure of their entrance and prayer hall, as shall be explained below.

The source for placing the entrance in the wall facing Jerusalem is found in talmudic commentary on praying in the direction of Jeru-

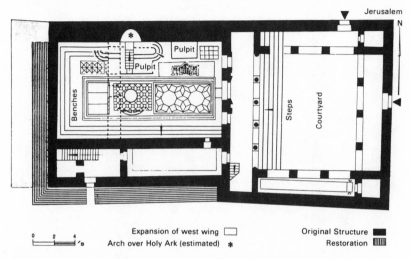

**Figure 19–5.** Plan of the synagogue at Khirbet Susiya. From *Qadmoniot* 5, 1972, p. 47. Used by permission of the Israel Exploration Society.

salem. This custom originates from Daniel:[1] "Rabbi Hiya bar Abba said Rabbi Yohanan said: *One should only pray in a house that has windows, as it is said* (Daniel 6:11) 'now he had *windows in his chamber open toward Jerusalem*, he kneeled upon his knees three times a day, and prayed, and gave thanks before his God, as he did aforetime'" (*Berakhot* 34b).

Several explanations given to this *halakhah* are connected to *kavanah* (intention) in prayer. The Rambam in *Hilkhot Tefillah* establishes the law that one should pray in the direction of Jerusalem. Then he adds (ibid., 5:6): "How is the place set? . . . *One should open windows or doors facing Jerusalem in order to pray facing it*, as it is said. . . ." The *Shulhan Arukh* in *Hilkhot Tefillah* 90:4 cites the Rambam. In *Hilkhot Beit Ha-kenesset* this *halakhah* is not mentioned as pertaining to laws of the synagogue, and section 150:5 of the *Shulhan Arukh* says that the entrance to a synagogue should be in the wall opposite the Ark.[2] However, apparently in earlier times the precedent from Daniel was interpreted literally, thus placing the synagogue entrance in the wall facing Jerusalem (see Figure 19–6). In certain ancient synagogues it was found that after an initial period of use, doorways facing Jerusalem were walled up and other entrances were made in the opposite

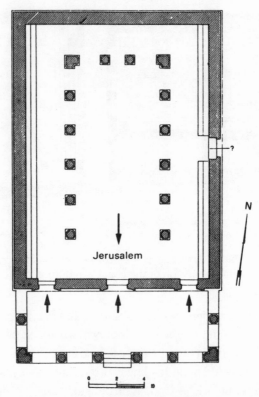

**Figure 19–6.** Plan of the synagogue at Baram. From *Encyclopedia of Archeological Excavations* (in English), ed. Michael Avi Yonah (Jerusalem: Israel Exploration Society, 1975). See "Kefar Bir'am" in Part III, p. 707. Used by permission of the Israel Exploration Society.

wall (Beit She'arim, Hammat Tiberias, Meroth) or in a side wall (Ein Gedi).

In this group of ancient synagogues the facade facing Jerusalem was beautified by stone sculpting and built with three entrances—a large central opening with two smaller openings at its sides (see Figure 19–7). This structure resembles both the entrance to the Temple Sanctuary and the Nicanor Gate that stood on the border of the Temple courtyard, in the east, leading from the *Ezrat Nashim* (Women's Court) to the *Ezrat Yisrael* (Court of the Israelites). According to the *Mishnah*, they were triple gates, the central one of which

**Figure 19–7.** Baram synagogue—the front, facing Jerusalem. Photo by Asher Grossberg.

was the largest (see Figures 19–8 and 19–9): "The height of the entrance to the Temple was twenty cubits and its width ten cubits . . . *and the large gate was flanked by two small gates: one in the north and one in the south (Mishnah, Midot* 4:1–2) . . . and in the east was the *Nicanor Gate. It was flanked by two small gates, one at the right and one at the left"* (*Mishnah, Midot* 2:6).

## SYNAGOGUE STRUCTURE AND THE STRUCTURE OF THE *EZRAT NASHIM* IN THE TEMPLE

The ancient synagogues in the Gallil are similar to the Temple not only in the structure of their entrances but also in the structure of their interiors. In these synagogues a row of columns was built parallel to every wall except the southern one. As has been said above, this wall faces Jerusalem, and in it were the entrances to the synagogue. Only in a few of these synagogues was the northern wall also free of columns. There is archeological evidence in some of these synagogues

**Figure 19–8.** Baram synagogue interior, looking in the direction of praying toward Jerusalem. Photo by Asher Grossberg.

of an upper gallery built above the aisles (see Figures 19–10 and 19–11).

Researchers attribute the architectural source of these synagogues to structures such as the Roman basilica, the courtyards in front of the pagan temples in Transjordan and Syria, and Herod's guest rooms in Jericho.[3] However, it appears that the inspiration for the internal form of the synagogue and its facade was drawn from the Women's Court, the Nicanor Gate, and the sanctuary entrance of the Temple. This is not to deny contemporary architectural and decorative influence on the ancient synagogue which makes it externally similar to other buildings of the same period. However, as will be clarified further on, the synagogue was different in essence from the pagan temples built in the same period. The inspiration for its unique structure came from no place other than the Temple.

The Women's Court was situated in front of the Nicanor Gate (see Figure 19–10). Its dimensions were 135 × 135 cubits. In each of its four corners was a 40 × 40 cubit chamber. Thus, the area adjacent

**Figure 19–9.** Meiron, restoration of the synagogue interior (3rd century C.E.). The facade of the building faces south—to Jerusalem. From E. Meyers, *Excavation at Ancient Meiron, Upper Galilee 1971–72, 1974–75* (Cambridge, MA: American Schools of Oriental Research, 1981), p. 13, fig. 2.5. Used by permission of the American Schools of Oriental Research.

to the Nicanor Gate was only 55 cubits wide. The *Mishnah* relates that a gallery the length of the walls of the Women's Court was built for the women first, especially for the Water Drawing Ceremony festivities of Sukkot:[4] "Previously it was a smooth wall and [later] they surrounded it with a gallery *so that the women should look on from above and the men from below in order not to be intermingled*" (*Mishnah, Midot* 2:5). The *Tosefta* adds that the structure of the gallery was only the length of three of the walls of the Women's Court: "*They made three galleries in the Ezrat Nashim along the three sides, where women sat and watched the Water Drawing Ceremony*" (*Tosefta, Sukkah* 4:1).

It is reasonable to deduce that the one wall upon which a gallery was not built was the western wall containing the Nicanor Gate (see

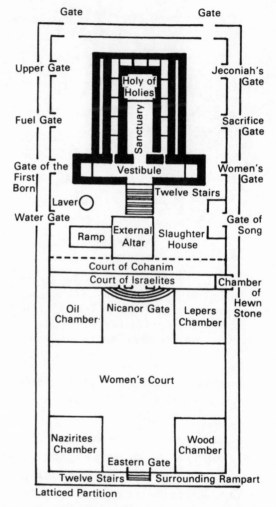

**Figure 19-10.** The Temple and its courtyards. From A. Albeck, *Mishnah, Seder Kodashim*, tractate *Midot* (Jerusalem: The Bialik Institute, 1958). Used by permission of the Bialik Institute.

**Figure 19–11.** The Nicanor Gate and the front of the Temple, looking from the Women's Court. (Postcard of the reconstruction model of the Temple at the Holyland Hotel, Jerusalem, used here with permission of the hotel.) From *A Short Guide to the Model of Ancient Jerusalem* (Jerusalem: Holyland Hotel).

Figure 19–11). During the rest of the year the *Ezrat Nashim* was not used exclusively for women. All pilgrims had to pass through the *Ezrat Nashim* to enter the *Ezrat Yisrael* on the other side of the Nicanor Gate. On certain occasions the *Ezrat Nashim* served as a collecting point for all pilgrims or as the setting for a special event:

1. On Yom Kippur after completing his service in the Holy of Holies, the High Priest went down to the Women's Court and there read from the Torah to all the people.[5]
2. Once every seven years, on the Sukkot festival following a *shemitah* year (sabbatical year during which the land is not worked and debts are cancelled), the people gathered in the *Ezrat Nashim* to carry out the *mitzvah* of *hak'hel* as commanded in Deuteronomy 31:10–13. A wooden platform was raised there upon which the king read

from the Torah to all the people—men, women, and children (Mishnah, Sotah 7–8).

The ancient Gallilean synagogues had sanctuaries similar to the Ezrat Nashim in the Temple in certain ways. Most of them had pillars along the three sides. The southern entrance facing Jerusalem, towards which the congregants prayed, did not have pillars. The Ezrat Nashim of the Temple also had pillars along the three walls enclosing the wall in which stood the Nicanor Gate, which the people faced as they entered the Ezrat Yisrael. In both, the people saw in front of them a triple gate: two small entrances flanking a larger one.[6] They were structurally dissimilar in that the Gallilean synagogues were roofed whereas the Ezrat Nashim of the Temple was not.

As we said above, all the pilgrims passed through the Ezrat Nashim on their way to the Ezrat Yisrael, and the separation between the men and the women was made on Sukkot, during the Water Drawing Ceremony. It is reasonable to assume that in other large public gatherings—such as the two public Torah readings—such a separation was also made.

In some of the ancient synagogues, the archeological evidence points to an upper gallery built on top of the aisles created by the columns. This may be interpreted as the place of the Ezrat Nashim in the synagogue, which is like the Ezrat Nashim in the Temple. Talmudic sources support this hypothesis.[7]

The Tosefta (Sukkah 4:4) and the Babylonian Talmud (Sukkah 51b) say that the synagogue in Alexandria was "like a large basilica, with colonnade within colonnade." That is to say, it was a building with rows of double columns. The Jerusalem Talmud (Sukkah 5:1) describes its destruction by Trajan the Wicked, the Roman Emperor Marcus Ulpius Trajanus, against whom diaspora Jews rebelled from 115 to 117 C.E. The Talmud says that the widowed Jewish women of Alexandria dared Trajan, "That which you did [to the men who sat] below, do also [to the women sitting] above [on top of the colonnades]."[8]

Synagogues are constructed with their interiors facing Jerusalem and the Temple. This is because not only must the worshipper face Jerusalem and open apertures towards it—as was done in the ancient synagogues—but he must also feel as if he were standing in the Temple (*Yevamot* 105b; *Shulhan Arukh, Hilkhot Tefillah* 95:2).[9] The inspiration for the structure was drawn from the same place—from the Temple and its *Ezrat Nashim* where the Torah was read every year on Yom Kippur and every seventh year on Sukkot to perform the *mitzvah* of *hak'hel.*

The synagogue serves as a place to pray and read the Torah in public. Public reading of the Torah was established before public prayer. Attributed by the Talmud Sages to Moses and Ezra,[10] the custom of reading the Torah in public can be seen as the initial reason for having synagogues.[11] The structure of the synagogue was chiefly influenced by the arrangement of the *Ezrat Nashim* of the Temple where the Torah was read in public on special occasions and through which everyone had to pass to enter the *Ezrat Yisrael.* The need for a separate place for women in the synagogue was met, then, by allotting them a gallery like the one in the Women's Court of the Temple, where they sat during special events.

As Daniel had done in his time, the ancient synagogue congregant prayed facing an aperture that opened toward Jerusalem. The triple form of the large door flanked by two smaller ones together with the rows of pillars lining the three surrounding sides reminded him of the sight revealed to every pilgrim going through the *Ezrat Nashim* to the Nicanor Gate to the *Ezrat Yisrael.* This vision apparently later guided synagogue builders to construct "small temples" that reminded the people of the destroyed Temple about which they had heard from their fathers (see Figure 19–12).

As has been said, the ancient synagogue builders drew their inspiration from the Temple while also borrowing structural and decorative elements from contemporary architectural traditions and adapting them to the special needs of the synagogue. Moreover, near the Temple itself there had been two basilica structures: the Chamber of Hewn Stone where the Sanhedrin held court—which was near the Nicanor Gate in the *Ezrat Yisrael*—and the royal porticos which stood on the south of the Temple Mount.[12]

The differences between a civilian Roman basilica, the palace basilica and the early Gallilean synagogues lie in the number of rows

**Figure 19–12.**   Ancient synagogue at Meroth, layers 1, 2 (5th–6th centuries). The front of the synagogue faces south—to Jerusalem. From T. Ilan, "Hidden Treasures in the Gallil: Meroth—an Ancient Jewish Settlement," *Teva va-Aretz*, vol. 25, no. 5 (July–August 1986): supplementary illustration. Used by permission of *Teva va-Aretz*.

of pillars parallel to the walls. In a civilian basilica there were four rows of pillars parallel to the four walls and the roof of the central part was elevated. In the palace basilica there were two rows of pillars running the length of the hall or there were no pillars at all, whereas in most of the synagogues there were three rows of pillars running parallel to three of the walls. The front side of the synagogue had no pillars, and a gallery was built on top of the three rows of pillars.

Although the facade of the synagogue resembled that of the pagan temple in its ornateness, it differed greatly in its significance. As the interior of the pagan temple in general was reserved for statues of gods, its decorative facade was very important because the public watched from the outside courtyard without entering inside. The facade of the synagogue served as the entrance for the public, and from inside the interior hall the congregants turned toward the open doors to pray facing Jerusalem. The Torah, however, was read in the center of the synagogue. It was only during the reading of the Torah that the center of the synagogue was the focus of activity. The synagogue congregation participated in the activity conducted in the interior, as opposed to the pagan worshippers, who stood outside watching passively. Whereas the idolators stood outside looking inside the temple, the synagogue congregants inside the synagogue stood facing outside, through the open doors, toward Jerusalem, toward the Temple, thinking themselves there.

## THE PHILOSOPHICAL MEANING OF THE *HALAKHAH*

The Bach, cited above, refers to synagogues with vestibules in front of them. He attributes a deeper meaning to the vestibule and the *halakhah* discussed at the outset of this article. This meaning distinguishes between synagogues and pagan temples and supports the need to build a vestibule in front of the synagogue. The Bach explains that prayer to the Holy One Be Blessed is done without any intermediary, and therefore the worshipper must enter into the heart of the synagogue, and pray there—not remaining outside in the vestibule. In contrast to this, the sanctuary in temples of idolators was not made for the public but for a statue, while the public remained outside.

As we also showed above, this difference that the Bach points to can be seen in those ancient synagogues which do not have a vestibule in front of the entrance. In these synagogues the wall facing Jerusalem had three entrances which, on the one hand, were externally similar to structures found in other kinds of buildings but, on the other hand, had a unique spiritual essence.

Let us cite the Bach's own words:

> It seems that the reason is that . . . when one wishes to make a request from a king, one does not actually personally enter the king's private chamber. Instead, one stands outside in the courtyard and asks one of the king's servants to deliver the request. But in the case of the King of Kings it is forbidden to petition through an intermediary, neither through an angel nor a seraph. And because of this principle one must enter inside through two

doors. That is to say that even though I have come to the vestibule leading into the private chamber of the King of Kings, the Holy One Be Blessed, I do not engage one of His servants to enter into His chamber and deliver my request, but rather I enter into the synagogue itself, into the chamber of the King of Kings, the Holy One Be Blessed, and petition Him directly and not through any intermediary. (*Tur, Orah Hayyim, Hilkhot Tefillah* 90)

From *B'Or Ha'Torah* 6E, 1987. Translated from the Hebrew by Ilana Coven, with thanks to Eliot Braun and Ditsa Eshed of Jerusalem.

# History

# 20
# Worlds of Difference

*Yoseph Udelson*

Over the past five hundred years Western civilization has discredited, in turn, first the spiritual, then the intellectual, and finally the aesthetic aspects of the human personality. Only the mundane dimension of life remains intact for our question. The Jew, however, has an integrated multidimensional alternative to this monolithic reality.

Only recently have I come to appreciate how very great is the gulf between the Torah world and that of contemporary Western civilization. The difference between the two is not merely one of nuance, ideology, or even contrasting paradigms of thought. Rather the distinction embraces the very nature of the opportunities presented to a person through which he can actualize and expand his innate potential. In other words, it is not a case of competing weltanschauungs but—literally—of entirely separate worlds in which a Jew chooses to develop. What he can think, say, and do with himself depends on in which of these worlds—Torah or secular Western—he selects to live. These two worlds are not equal; they do not at all provide the same range of opportunities for the growth of human potential.

257

It was while teaching a university seminar on European intellectual history that I first became aware of the nature of this distinction. My students were very bright. I found the course stimulating but also disturbing. Throughout the course, my twelve students expressed radical cynicism whenever the discussion focused on the human personality and its potential. They jeered at the optimism of nineteenth-century political philosophers and rejected strongly the arguments of reform, betterment, and human nobility. The very notion that human beings are more than brutish, self-serving, pleasure-seeking creatures aroused their derision.

The hopelessness and cynicism of these bright young people struck me. Theirs was not the considered response to a world that has witnessed the destructions of Auschwitz and so rejects the illusory nostrum of Tolstoyan pacifism. Indeed, it was merely the everyday, unchallenged, received wisdom of intelligent young Americans: there is no hope, so learn to play the game well, know all the tricks, and at least be a winner.

And they are right! These bright young people really have penetrated through the last glow of past illusion and know their world to its core. Their radical cynicism is merely the intelligent verbalization of the same desperation others express through drugs, compulsive sexual promiscuity, the punk look, and video games. Nowhere do these people see anything higher than the mundane, so the mundane is itself elevated and worshipped. The intelligent among them know that they are worshipping an idol of stone that neither thinks nor acts, but they know of no alternative.

## THE TWO FOCI OF THE MUNDANE

In this world of the mundane there are but two foci: economic advance and pleasurable entertainment. These two foci are really opposite sides of the same coin because the mundane is concerned with the physical needs of the body, its survival and comfort in a supportive environment. These are perfectly reasonable goals, as far as they go. But the human personality contains higher, more refined levels of endeavor: the aesthetic, the intellectual, and the spiritual.

Throughout the world, over the course of history, humans have displayed a powerful need for the aesthetic. Thirst for beauty has found extraordinary expression in the fine arts, literature, music, and dance. Paralleling this, humans have equally evinced a hunger of the intellect, a drive to understand and comprehend a wide variety of phenomena, ranging from the most concrete to the thoroughly abstract. Equally impressive is the history of man's quest to please and propitiate the Power he intuitively feels transcends him.

Today, however, the mundane predominates. The intellect is scorned by the new Luddites, terrified over every advance announced in the sciences and technology, seeking instead the cuddly security of motion picture fantasy characters, and mouthing Rousseauist clichés about "caring," "feeling," and "touching," the "natural" and the "spontaneous." As intellect is scorned, education has foundered in a morass of directionless experimentation with "relevance" (i. e., mundane concerns); political and ethical theorizing has been reduced to knee-jerk responses to code words devoid of content. Even philosophic thought has been reduced to irrelevant arid and arcane discussion.

For all the "caring" and "feeling," there is no art. The emotive is drained into the frivolous and lost to the creative. The fine arts, literature, serious music, and dance languish amid the aesthetic triumphs of past generations. Artists, authors, and composers are no longer held in high public esteem but are now merely government pensioners, university lecturers, and television personalities. Nor is this merely an expression of disillusionment of the type that produced the immensely talented and productive "lost generation" of artists, authors, and composers after World War I. Rather, our current situation is the natural result of the devaluation of all concerns not obviously relevant to the service of the mundane.

Similarly, religion in contemporary Western civilization no longer focuses upon the transcendent. Now religious leaders are expected to be exponents of sociopolitical special interest causes competing for the attention of the mass media and the favors of the government. Theology has been reduced to sloganeering, and references to the soul have become a source of immense embarrassment.

## WHEN SERIOUS THOUGHT
## IS DEEPLY SUSPECT . . .

In the totalitarian East, Communist party authorities have attempted to eliminate the private domain by making every realm of the individual's existence a concern of the state. The resulting aridity of life in such circumstances has long been noted. But the same politicization of life is also becoming a marked characteristic of Western civilization. And it is no wonder; politics—the administration of the economic relations in society and the legitimization of its forms of entertainment—is the social expression of the mundane. So when only the mundane is esteemed there is a natural tendency toward the politicization of people's lives, as well as towards the impoverishment of politics. Intellectuals are accorded attention by the public now only when they can successfully compete with photogenic entertainers as political pundits. Culture is funded only when it can demonstrate its political relevance or its mass entertainment value. Religious spokemen are taken seriously only when issuing political statements couched in moralistic jargon. And in a period when serious thought is deeply suspect, politics itself has been reduced to policy by media events, its effects judged by the entertainment value of the news "story."

## WESTERN CIVILIZATION HAS
## NOT ALWAYS BEEN LIKE THIS

No wonder my students were so cynical. This is the only condition of the Western world they have ever known. But from the historical perspective they are wrong. Western civilization has not always been like this. It has known long eras of the most optimistic strivings after the highest levels of human potential. It is actually only for the past two decades that Western civilization has been so constricted within the mundane.

For instance, at the culmination of the medieval period of Western civilization, in the fourteenth century, the spiritual was accorded exclusive legitimacy. The intellectuals of this era, the natural philosophers, concentrated on exploring the nature of the heavens and hells and the glories of Divinity found on Earth. The fine arts, literature, and music served only to heighten the impact of the spiritual realm. The individual and his mundane struggle to survive were haughtily despised as too ephemeral, and hence too trivial, to be allowed to detract from attention to the spiritual.

Such psychological constriction could not long continue. The Renaissance of the fifteenth century was characterized by a violent, often petty, individualism and by an impressive flowering of intellectual and aesthetic creativity. These new impulses represented a rebellion of the secular against the spiritual. But it was only after the prolonged and bloody wars of the Reformation, generally over by the seventeenth century, that the spiritual was finally tamed by the secular.

Pride of place was now accorded exclusively to the intellect. The Enlightenment of the seventeenth and eighteenth centuries witnessed the birth of modern science, culminating in the Newtonian synthesis, significant advances in the life sciences, and the technological achievements which provided the impetus for the Industrial Revolution. At the same time, Cartesian rationalism and Lockeian environmentalist associational psychology created the expectation that human reason alone was sufficient to engineer a progressive society free of ignorance, superstition, corruption, and suffering. Enlightenment intellectuals posited transcendence, spirituality, and aesthetics as their "enemy," while they continued to ignore the mundane realm as unreliable and trivial.

But these great expectations were shattered in the bloody irrational violence that accompanied the French Revolution and its attempts to conquer Europe. As earlier devastations of the wars between opposing religions destroyed the preeminence of the spiritual in Western civilization, so now the destructiveness of the French revolutionary crusade discredited reliance on intellect. In the aftermath of Napoleon came Romanticism and the elevation of the aesthetic as the sole legitimate realm for the development of human potential.

The nineteenth century witnessed immense and impressive cultural productivity across the entire spectrum of the arts. At the same time, the aesthetic, with its stress on the organic, the emotive, the historical, and the particular also finally accorded attention to the mundane in Western civilization. Physical existence was viewed as the basis of humanity. For many researchers this meant adopting Comte's program of Positivism. In the arts it led to a readiness to experiment in attempting to portray an inner reality lying behind surface appearances. And in politics, it meant, on the one hand, concrete socio-economic reform and, on the other, the rise of romantic revolutionary movements such as *volkische* nationalism, Marxism, and anarchism.

The aesthetic continued to dominate Western culture until the end of World War II. By then, creativity in the arts had generally become exhausted and had sunk into a sterile experimentation with abstract form. Hitler and Stalin succeeded in thoroughly discrediting the aesthetic in Western politics. And so, like its two predecessors, the aesthetic was discredited in a holocaust of destruction, the most horrifying in human history.

## NOTHING LEFT TO DISCREDIT
## BUT THE MUNDANE

The post–World War II generation grew to maturity amid the rubble of the shattered illusions of over five hundred years of Western thought. The spiritual, the intellectual, and the aesthetic had each failed to provide the proper arena for a full and healthy development of the human personality and, consequently, each had been discredited amid the violence and carnage that erupted out of its failure. Emerging from this wreckage, the post–World War II generation could find only the mundane intact. Thus it is that day-to-day physical existence, so long despised as a necessary but trivial contingency of human life and treated only as a means of transcendence by the Romantics, has become the only goal of human endeavor granted legitimacy by Western civilization.

Mankind is reduced to the animal. But of course such a tremendous constriction of human potential will not long be tolerable: it too greatly violates the structure of the human personality. In fact, it is not even working now. A world providing such a confining range of opportunities for human development cannot absorb and satisfy the energies and desires of the personality. There is a continual disquiet about us and a rebellion against the "existing order."

But this constant seeking of "alternatives" is limited generally to experimentation within the limits of the mundane: new modes of manipulating physical appearance, new stimulants, new fantasies of infancy and escape, new sports leagues and playing fields, new careers as refuges from "job burn-out," new corporate mergers, new political causes to right newly discovered wrongs.

The vast realms of human potentiality beyond the mundane have been forgotten or are remembered now only as bad dreams. Hence my students' incredulity and cynicism. Their world, the Western world, now provides legitimate opportunities only within the narrow span of employment and entertainment. No wonder the range of what is considered acceptable within this domain, and particularly what constitutes "entertainment," constantly is being expanded. The very education necessary to find alternatives to the constrictions of this world is no longer available, to say nothing of social tolerance of even modest professional opportunity. And so anguished disquiet, frenzied search for change and continuous rebellion proceed unchecked by any prospect of satisfaction.

However, I do not believe the Western world is by any means doomed to this situation forever. This would require too great a violation of basic psychological needs. Being an historian accustomed to scrutinizing the past, I am leery of practicing futurology. So while I think change will come, I cannot guess whether it will be through some new creative surge back into the more refined levels of the personality, the establishment of the benevolent totalitarianism of some drugged "brave new world," or conquest of a moribund civilization once again by a new horde of barbarians, or some other course, until the soon expected coming of the Messiah.

## THE VIBRANT ALTERNATIVE

For the Jew there already exists a genuine alternative, a vibrant world exactly molded to his personality: the world of Torah. And it is no wonder; the Creator of the human personality is the same Author of the Torah in which it should be clothed. Unlike Western civilization (the finite product of finite human minds inevitably limited by historical and psychological myopia), the world of Torah is the creation of the Infinite One Who grasps exactly the requirements of each of us. Where Western civilization has always been too constricted to meet the needs of even the most simple person, the Torah is spacious enough to embrace the needs of everyone.

In the world of Torah, every element of creation is satisfied. No psychological level is excluded, and none improperly emphasized. For instance, on the mundane level of action, daily concerns are neither too trivial to be of any account nor merely artifacts out of which some private inner life may be constructed. Rather, they are important for themselves, tasks to be elevated in the service of God. So one's daily life is either the pursuit of Torah and its Commandments or the means for such service, thereby itself becoming elevated. The satisfaction of physical needs becomes a vehicle for sanctifying the body. Even recreation refreshes and invigorates a person so that he can continue his Divine service with renewed energy and vitality.

Above the level of daily activities lies the creative and emotive realm of the personality: the aesthetic. Here, too, there is a broad range of possibilities open to every individual, according to his inclinations and aptitudes. Art and photographic studies have come to occupy a significant place in the Torah world, often appealing to emotions more subtle than words can reach.

Similarly, *nigunim* and *zemirot*—melodies and songs—employ sound to affect the soul in a manner often beyond the capacity of prosaic speech. Dance uses the limbs of the body in a similar way. The hasidic tale and the children's story can touch our imagination and thought sometimes more easily than the reasoning of didactic exposition. The aesthetic thus operates as complement, refinement, and spur to the mundane.

Higher than the emotive level of the aesthetic is the domain of the intellect, on both the levels of the *nigleh* (revealed wisdom), the Oral Teaching, *halakhah* (the Law), and speculative philosophy, and of *nistar* (hidden wisdom), the *Kabbalah*, and Hasidism. The age-old Jewish tradition of learning from childhood onward is not by any means training to earn an income. It is not even an exercise of the intellect for its own sake. Rather, Torah learning is a profound and vigorous means of serving God and elevating the soul. It is, as well, an inspiration and necessary guide to proper service on the aesthetic and mundane levels.

Highest of all is the desire to cleave to God Himself. But this spiritual level of the personality is not a distinct compartment separated from the other levels. Instead, it is the inner longing and intention we strive to bring to every one of our thoughts, words, and deeds in the world of Torah service.

With every action and every second possessing the potential to serve as a vehicle for holiness, in the Torah world there is no time for cynicism. There are so many important opportunities waiting and we have so much to do.

From *B'Or Ha'Torah* 5E, 1986.

# Music

# 21
# Where the Roads Meet: Composer André Hajdu's Musical and Jewish Identity

*Yaffa Goldstein*

If ten measures of wandering were apportioned to the world, it would seem that most of them were taken by the Gypsies and the Jews. Such a wanderer was the man who sat before me one rainy Jerusalem evening and spoke, with eyes closed, of his three years with the Gypsies, of his many years of wandering, of the gradual discovery of his Jewish roots, and of his music.

I search his face for traces of those wanderings, and hear that all the diverse and circuitous paths have now merged into one unique road. My glance combs the living room of his modest apartment and stops at two prominent foci: the black piano and the table covered by holy books. Two main signposts of the road on which composer André Hajdu now travels.

Hajdu is a thick-bearded man, and an ample skullcap crowns his head. Yet even if one were an expert on the various factions within Jerusalem's Torah-observant community, one would have difficulty identifying him with any particular group.

"When did music and Judaism begin to dwell together in my world?" Hajdu furrows his brow and plunges into a sea of memories. "Music has almost always been with me." Hajdu began playing piano as a boy. Later he studied composition and sensed that his future lay there.

His Jewish side, however, was undeveloped. "I was born in Budapest in 1932. My parents' house was a bourgeois, assimilated home, something like the average assimilation in Tel Aviv. In Hungary there was a gap between religious and secular Jews." Jewish memories of his childhood are dim; the most enduring is of a Passover *seder* at his uncle's home. The taste of the different foods, the aromas, the spirit and song of that *seder* night remain with him—but there was no continuation. Nor was there any connection with Jewish nationalism or Zionism. Hajdu belonged to a circle of cosmopolitan intellectuals that vigorously opposed communism until the outbreak of the Hungarian Revolution and its subsequent Russian repression.

## THE GYPSY PERIOD

The person who spoke most to Hajdu's heart was the composer Zoltan Kodaly, who sought to discover the authentic folklore of the people, to find succor in the nation's roots, for only through the revelation of this truth could fine classical music be created.

"And so," explains Hajdu, "I was introduced to a band of wandering Gypsies—simple folk, horse traders. They were genuine Gypsies. I was then very distant from the Jew within me. I became very involved with them and learned their language so well that even the Gypsies themselves mistook me for one of them. Without my being aware of it, this was really a search for roots, for my own source."

These wanderings gave birth to Hajdu's first symphonic composition, "Gypsy Cantata," which debuted in 1955 and won first prize in the Warsaw Youth Festival.

After the repression of the Hungarian Revolution in 1956 and the temporary opening of the borders, Hajdu felt intuitively drawn to the West. Indeed, the West had always attracted him, as a cradle of great culture that never reached Hungary.

## THE FIRST PARISIAN PERIOD—
## CULTURAL WEALTH

Hajdu arrived in Paris. This metropolis, so beautiful and open, quenched his thirst for the books and films he never saw in Hungary. "My first three years in Paris introduced me to things that broadened my world, which was of course enhanced by that special flavor of a Bohemian life-style, where one never has to worry about the future. . . ." Only later came the reservations, the concern about becoming mired in that Bohemian life, a life in which one need not create anything—in which being free to socialize is one's sole obligation. It was then that existential questions began to arise. Hajdu had the choice to look for the answers in one of two places: in France, where he could research its Gypsies, or Tunisia. It was very difficult then to become part of the Paris music world, and it seems that without knowing it, Hajdu was already looking for different answers. Thus he found himself accepting a teaching position in Tunisia.

## TUNISIA—THE EXISTENTIAL PERIOD

If Paris had erased some of Hungary's influence, in terms of education and values, Tunisia was a continuation of this process. Hajdu describes it thus: "There was much less movement in Tunisia. Much less art. Neither was there any push to become famous. On the other hand, I received something deep, in terms of basic, essential experience. From the distance of Tunisia my three years in Paris seemed like a series of pictures on a screen, totally unconnected to me, as if none of it had ever happened to me. I tried to experience every moment with intensity. There I discovered myself as a human being and not just an artist. I found that I could build something stable and lasting even without those things that I needed as an artist in Hungary and Paris."

In Tunisia Hajdu became friends with a Jewish family. Together they travelled to Djerba. The encounter with the remnants of the island's Jewish community, the similarity between the Jewish houses and those of his Eastern European village, the Jewish cemetery and synagogue—all were a provocative experience for Hajdu.

His greatest discovery was the strong common bond with his Tunisian Jewish friends. "I found that despite the differences in our thinking and way of life, they were not really Arabs or Frenchmen, just as I was neither Hungarian nor French. I felt that we were part of the same people. Until then I had lacked the key. I saw the Jew in Hungary as a Hungarian of the Mosaic persuasion and did not understand what could connect him with a Jew from another country. Only in Tunisia did it become clear to me what I had really been searching for all those years in Hungary and Paris."

Hajdu's spirit flourished, but the musician in him languished. During his two-year sojourn in Tunisia he composed nothing. After two years he returned to Paris with a new awareness. "I returned not as a Hungarian coming to Paris to pursue an artistic career, to be a Parisian, or a cosmopolitan, but as a Jew."

Hajdu was not yet an observant Jew, but he did not return to his international environment. His Tunisian friends also came to Paris, and Hajdu found work conducting a children's choir in a Jewish orphanage near Paris.

## THE TURNING POINT

The environment in which Hajdu now lived was indeed Jewish, but not religious. On Passover *seder* night that year Hajdu was a guest at the home of a Jewish friend whose father, Yosef Gottfarstein, was well versed in Jewish philosophy, though not observant. His explanations of the *haggadah* and its symbols aroused great interest in Hajdu. After Passover he began to study Hebrew with Gottfarstein. At this point Hajdu was spiritually ready to make a far-reaching change in his life.

The turning point was a renewed acquaintance with a film director, who invited him to write the score for a film he was making in Rome and Greece. The friendship between the director and the musician deepened. The director was an observant Jew, and when the rest of the film crew would go to the beach on Saturday, Hajdu would accompany his friend on his long walk to a kosher Jewish student cafeteria. On one of those Sabbaths the director suggested that they study Talmud together. "It was a complete surprise," recalls Hajdu. "Before my eyes a new world opened in which legal thought was integrated into a melodic folk idiom." The words had such a natural melody that he wrote it down as a way to remember what he had learned.

With the completion of the film, Hajdu returned to Paris and began studying Talmud regularly with a rabbi. In the beginning it was purely an intellectual exercise, and a difficult one at that. After leaving a Talmud lesson Hajdu would remove his skullcap and eat nonkosher food.

After a short time he was introduced to a group of young, newly observant intellectuals. "We all had something in common, whether intellectually or artistically." These new friends helped Hajdu shed his negative image of the religious Jew.

Gradually, Hajdu began to observe *mitzvot* (commandments), a development he believes could only have happened to him in the West. "In the Hungary of Eastern Europe, Judaism seemed too Eastern, too primitive. I wanted to escape from it. I was drawn to the West."

It was only after he had encountered the wealth of Western culture that Hajdu discovered the depth of Judaism in the West. Without this exposure to the strength of the West, which served as a kind of yardstick for Judaism, Hajdu might not have embarked on his path of intense introspection and his search for roots.

## MUSIC AND THE RETURN TO JUDAISM

Not long thereafter Hajdu was offered work researching hasidic music at the National Institute for the Research of Jewish Music at the Hebrew University in Jerusalem. Hajdu made *aliyah* (immigrated to Israel) and had few difficulties adjusting to his new environment. "My greatest interest was the ethnography of the Jewish world, with all its communities, and the musical research was just a way for me to seek out the roots of my people, just as my research of Gypsy music had been a means to discover the roots of that strange tribe." However, in Israel Hajdu did not conduct ordinary research on some exotic sect in some faraway land. Here a work had meaning not necessarily because of its musical quality but by dint of its identification with the people from whom it had emerged. Hajdu explains, "I wasn't looking for an all-inclusive framework for all the different types of Jewish music. I was looking for the kinds of people and special communities I had not yet encountered, those through which I could better sense the pulse of the people of Israel. That's why I didn't deal with urban forms but was driven more to the exotic extreme, for example, to the music of the *hasidim* of Rebbe Nachman of Braslav."

It was apparently not by chance that in Israel he married and here his six children were born.

## A CONTROVERSIAL COMPOSITION

Anyone who expects to hear that in Israel Hajdu's life came full circle and henceforth flowed smoothly within a defined framework is in for a surprise: the milestones of this composer's career, from the time he wrote his first work in Israel, have not marked a simple and straightforward path. That first piece, performed in 1977 and entitled "Passover Play," aroused considerable controversy within Israel's musical community. The script was based on medieval texts and described Christian children playing a game in which they pretend to be Jews. Carried away by the force of their scenario, they end up killing one of their own Christian playmates. Hajdu sought to make a powerful statement about the trauma of anti-Semitism. He relates, "I did not want to do this through the eyes of an adult who understands the meaning of evil. Children lack a clear distinction between good and evil. It seems to me that the cruelty stands out in even greater relief because it is seen from two perspectives at once: the brutality of the children and the suffering of the boy. This is an explosive treatment. People can't accept things that put them on both sides of the fence."

Hajdu's approach, which seeks to convey ideas in their raw, even chaotic form, inevitably explodes from the tension of its internal contradictions. In many ways it characterizes the path of the man often referred to as a "controversial composer." The seeds sown in Hungary and Paris continue to bear fruit in his life and work, but now Hajdu imbues them with Jewish content. In "Passover Play" one finds the influence of the school of Zoltan Kodaly in the emphasis on national roots and the use of folklore and children's songs. When Hajdu first came to Israel, he underwent a kind of return to his Hungarian period. In that spirit he composed, in cooperation with the clarinetist Giora Feidman, "Rhapsody on Jewish Themes," which incorporated the songs and melodies of various Jewish communities.

Today Hajdu does not feel the need to compose a work like "Passover Play." He has new goals, such as trying to understand the Torah through the perspective of the Israeli experience. The relationship between the false and true prophets in chapter 13 of Kings I is a case in point. In his composition "The False Prophet and the True Prophet" he seeks to portray the internal contradiction of the false prophet, who, though a prophet in every sense, chooses a path of falsehood and evil.

Until now, Israeli composers have treated the Bible in a way Hajdu describes as "historical" or "pedagogic." He thinks this approach is too detached in its pretense of "enlightenment" and thus misses the tugs of reality—in this case, the profoundly problematic nature of prophesy.

Hajdu also developed an idea inspired by his study of *Mishnah.* The result was a piece entitled *"Mishnayot,"* based on chapters of the talmudic tractate *Baba Kamma.* Hajdu worked on this composition for a year and a half with an enthusiasm and intensity unequalled in his own experience. The style of this composition was very similar to that of Hajdu's Parisian period: less folklore and more personal expression. He sought to express the drama hidden in every day-to-day situation described in the *Mishnah.*

*"Mishnayot"* also stirred controversy, perhaps because Hajdu did not employ "Jewish music" in the piece. He sensed that the *Mishnah* itself was sufficiently Jewish in spirit. The work was roundly attacked by Western music critics on the one hand and religious circles on the other. No one could accept his mixture of realms. The very attempt

to combine Western culture with the traditional study of holy texts aroused pointed indignation.

## WHAT IS JEWISH MUSIC?

Jewish music, according to André Hajdu, is "fabric created by Jews throughout the generations." Hajdu does not consider himself a composer representing Judaism, despite his reputation as "the religious Jew's composer." Although he has "not attempted to completely reconstruct Jewish music," he has tried to contend with many different aspects of Judaism: its folklore and history, as well as its psychological, allegorical, and educational dimensions. Yet in his portrayal of these dimensions there always has been an admixture of personal experience— coming from the universal elements of the childhood memory. Concerning this integration of Jewish and personal realms, Hajdu says, "Outside of Israel, I would not have been able to do this. In his time, Gustav Mahler felt he had to abandon Judaism to fulfill himself as a musician." Here in Israel, however, Hajdu has been able to create music that is religious by virtue of the fact that it deals with Jewish subjects, or because it echoes the musical voices of the Jewish world.

At the same time, Hajdu's music can be characterized as secular, based as it is on Western modes of expression and thought. A surprise was in store, for instance, for those who commissioned Hajdu to compose a piece based on Psalms, to be accompanied by a children's choir. Hajdu says that he intended to write a work that would be like "vinegar in the eyes." In other words, until then, Jewish liturgical composition was expected to imitate Christian forms. But Hajdu sought to present Psalms as an expression of the tension between the opposing worlds of his own soul, a tension comprising the diverse realities of human existence: suffering, transcendence, and in Hajdu's words, "all that I experience in the psalms."

Hajdu does not offer an interpretation of prayer per se, perhaps because of the associations he retains with church music, with its echoes of choir and organ. When asked whether music can play the same role in Judaism today that it did in the Temple service, Hajdu answers in the negative, explaining that we have no understanding of the music that was played then. "I am not an historian, and I don't deal in speculation. I'm interested in the present."

The present in which Hajdu lives is very complex, built on his complicated past and mirroring the complexity of secular and religious Jewish life in Israel today.

No longer is the line of demarcation between the two camps as clear as it once was. Factions of diverse hues abound on both sides. Still, the two blocs face each other in harsh confrontation, each claiming the other lacks real substance.

Characteristic musings of a Sabbath eve: Hajdu ponders a comparison of the clash between the secular and religious and the collision of the worlds of the child and the adult. The religious are accused of childishness and superstition, the secular of being cut off from the sources of spiritual emptiness. Hajdu believes that each side needs the other, just as adults and children are interdependent.

The religious camp is indeed spiritually mature but lags far behind the secular world in its civic development. Hajdu sees the relationship between the two worlds as a continuing dialectic, in which each side issues a triumphant response to the other, only to be challenged by rebuttal.

## A MEETING PLACE OF CONTRADICTIONS

Such is the world of composer André Hajdu. His personal life is a medley of voices streaming in from different directions: not only the voices of the religious and secular camps but also the distinct voices of the "enlightened" adult world and the mischievous world of the child, in which good and evil are blurred, and, not least, the voices of the world of tradition in juxtaposition to culture and art.

It is inevitable that someone standing in the midst of this cacophony would be a subject of considerable controversy. Those in the secular camp, enthralled with Western culture, see Hajdu as too parochial, too simplistic, too "kitschy." Kodaly's approach, in which the artist draws succor from the roots of tradition, is foreign to them.

The religious, on the other hand, accuse Hajdu of clothing holy texts in Western, symphonic garb and of disloyalty to the sources. In Hajdu's view, both sides end up losing—and the unwritten taboo that separates them places him in the eye of the storm.

"I don't do this intentionally to be different," says Hajdu. "It's like a river that flows from afar, from childhood to adulthood, from the world of tradition to the world of art. But the river in me doesn't flow in only one direction, for if it did, things would be much simpler. I flow in the opposite direction as well, from West to East, and many times from adulthood to childhood."

Hajdu is aware that he is a tangle of contradictions, whose source lies not in music but in the tension of the opposing forces of his personality. In music, this complexity is characteristic of avant-garde, which contends with two poles in order to portray the conflict between them. The only possibility of combining these opposing forces into one unified whole is to descend to the source of each one.

Can André Hajdu serve as a guide for others? It would seem very difficult to lead others through a course so tortuous, so full of contradictions. Hajdu recognizes this and has never tried to understate this challenge. But his very standing at the crossroads of so many opposing paths hints that somehow these contradictions are reconcilable, for at the end of the dialectical ladder awaits the final, absolute answer: the way of the Holy One, Blessed be He.

## AN AFTERTHOUGHT

André Hajdu's initial attempt at building a bridge between Judaism and music reminds me of the story of the fourth, bent-necked beggar in Rebbe Nahman of Braslav's "Tale of Seven Beggars." The story tells of a land plagued by terrible, frightening cries that disturb the sleep of all who dwell there and even bring the inhabitants to shriek uncontrollably themselves. Far from that land is another country that suffers the same woe. Their distraction drives the people to stray into each other's land.

Salvation for both places comes from the bent-necked beggar, who understands that the source of the problem is a pair of lovebirds who have become separated. Their longing for each other is so powerful that it bursts forth in a great cry that rocks the sleep of the denizens.

No man can approach the nests of these birds, but the bent-necked beggar is blessed with the wondrous ability to reproduce all the voices in the world. It is the very misshapenness of his neck that gives him this amazing vocal versatility. So the bent-necked beggar mentally records the cries coming from the two distant lands and then, with his miraculous voice, throws them back. Thus the connection between the two lovebirds is renewed and, through the sound of their voices, they and the people of the two lands are reunited.

Will André Hajdu bring together the voices calling forth from distant lands?

From *B'Or Ha'Torah* 7E, 1991. This article first appeared in Hebrew in *B'Or Ha'Torah* 2. It was translated into English by Yedidya Freeman. Since 1982, when the article was written, André Hajdu has composed *Opera on the Book of Jonah.*

## A RED INDIAN CALLED LEVY-STRAUSS: THE EVOLUTION OF JEWISHNESS IN THE COMPOSER'S MIND

*Excerpts from a speech given by André Hajdu at the opening of the Conference on Jewish Expression in Twentieth Century Music, held in New York, April 2, 1989.*

Jewishness—musical or otherwise—is a question that can be neither answered nor avoided. The existence of each individual Jew is actually the answer, most often without his being aware of its meaning. Usually, before this question is asked, the answer is already there, so strongly predetermined by education, environment, and other circumstances that I don't believe that even the most independent mind among us is really free to postulate it unconditionally. And when we verbalize it, things become worse. Most of us try to present the result of unknown impulses and influences as conscious choice.

\* \* \*

The polarity inherent in ethnology always seemed strange to me. On the one hand, there is the researcher—an isolated Westerner who must leave his natural milieu to come face to face with native existence. On the other hand, there is the autochthon, whose life has everything, but who is not conscious of it. The researcher lacks ethnic life, while the native lacks cognitive self-awareness. I always used to wonder why these two qualities could not coexist within one person or tribe.

At an ethnological seminar in Paris I once delivered a half-serious paper about this paradox. Everyone smiled, but nobody could figure out how things could be otherwise. Then later in Rome I found a key to coexistence. The very first page of *Mishnah* that I read in Rome gave me the clear impression—which has never since changed—that here is the perfect example of conscious autoethnology written by people who lived a folk existence, believed in it, and at the same time were able to understand and analyze it. The same "researchers" who consciously examined the laws, metaphysics, economy, and all the minutiae of Jewish life in the Talmud were at the same time popular heroes like Rabbi Akiba and Rabbi Shimon bar Yohai. Picture [the anthropologist] Levy-Strauss as a red Indian!

\* \* \*

Later, when I was already deeply involved in studying Talmud, I discovered that the way it is written and studied answers a basic need that I had been harboring since childhood. My parents used to tell me not to wave my hands while speaking and not to talk in a singsong. When I asked why, they said that it made me look like a Jew. It was in Israel that I first saw yeshiva Talmud study in action—with the forbidden hand motion and singsong. And then I understood how atavistic this behavior must have been in me.

At the same time, I began to discover that the entire Talmud— especially its basis, the *Mishnah*—is a poetical-musical construction as much as a legal and intellectual work. The Talmud is written in such an interrogatory way that one cannot help but sing it in an emotional manner with the inevitable hand gestures. . . .

My first years in Jerusalem were under a Bartok-Kodalyian constellation. My aim was to understand the language, music, and civilization of this manifold world and to become a part of it, while at the same time fulfilling the Hungarian composer's dream of researching and building a musical life based on ethnic truth. . . . If my life had stopped at that point, it would have been a beautiful and edifying construction, so perfectly fashioned that it might seem more like a musical composition than a human life. When I look back on it, I realize that my previous existence had been exactly what Germans call *Wanderjahre*, years of apprenticeship and journeys—unstable, not integrated into concrete life, lacking family and other ties. An idea could drive me then from one life-style to another, and I could change countries and continents as my convictions changed.

As you can probably guess, life in Israel was the exact opposite of all that. Not by chance the early Zionists called *aliyah* [immigration to Israel] *hagshamah* [realization]. *Hagshamah* also means incarnation. For me, it meant marriage, army, children, professional responsibilities. All these work in directions antithetical to "ideas." Curiously, though, this was the period of my self-realization as a composer. It was exactly what I wanted from Israel. I liked the heavier load and more concrete life. But I was sometimes disappointed by the narrowness of cultural options and the lack of capacity for complex thinking. Religiosity was reduced to petty right-wing politics, Western openness to a denial of Jewish roots. The syntheses, however, were usually worse, as happens when synthesis takes place on lower levels. . . .

# Psychology

# 22
# The Third Way: The Jewish Psychoanalytic Approach of Dr. Ida Akerman

## Sarah ben-Arza

When Dr. Ida Akerman's patients thank her for her help, they use the French verb *aimer* ("love") instead of *aider* ("help"). This slip of the tongue exposes the secret of Dr. Akerman's psychoanalytic treatment.

Dr. Ida Akerman is a tall, slender redhead with cognac-colored eyes, whose warm depth of sensitivity is revealed in her every expression and gesture. Her parents were Polish Jews who migrated from place to place in Europe and were killed in Auschwitz. In 1938 she moved from Germany to Belgium and, in 1940, from Belgium to France. After completing her studies in medical school in Paris in 1963, she started to work as a children's doctor. During this time she had three children. Her husband, also a doctor, heads the department of nuclear medicine at Fouche Hospital near Paris. The Akermans are Torah observant and active in the Paris Jewish community. Their two sons live in Israel. (One studied psychology at Bar Ilan University, and the other is preparing for the rabbinate after having studied electronics at the Jerusalem College of Technology.) Their daughter is a doctor in Paris.

Although Dr. Akerman feels as if she has always been a psychoanalyst, she officially started working in this field as a result of a discussion with a friend. The friend had consulted a psychologist to treat her son's physical and emotional problems. The psychologist was a Gentile, and in his analysis of the family's problems he ignored the mother's past—the past of a persecuted Jewish woman. In order to help her friend's son, she started a self-analysis through formal study of psychoanalysis.

## THE CHRISTIAN SPLIT AND
## THE MODERN FAMILY

At the center of Dr. Akerman's method is the concept of identity. "Identity is not only a collection of items found on an ID card," Dr. Akerman explains. She sees the adventure of life as a constant search for identity and its realization. Problems and complexes result from situations that make an individual betray himself and adopt alien identities.

Dr. Akerman demonstrates the problem of building a whole identity by contrasting the Jewish approach with the historical development of the Christian and the modern Western humanist outlooks.

Christianity divides man into body and soul. The body by necessity is bound up in sin, while the soul is pure. Christianity challenges man to deny his drives and inclinations and to concentrate on the purity of his soul. However, this challenge is neither natural nor healthy. Most devout Christians found themselves torn between their body and their soul. As it is impossible to ignore the body entirely, they were forced to connect soul and body by a bridge of hypocrisy. But hypocrisy is not a strong foundation upon which to build a personal identity. Indeed, this caused centuries of moral degeneration and corruption.

In trying to overthrow the Christian outlook, modern humanism went to the opposite extreme. In its zeal to overcompensate for what the Christian ideal had denied it, the Western world legitimized the almost unrestrained satisfaction of physical drives and desires.

The rise of psychology helped this revolution. This new science investigated the mechanisms controlling the soul. It studied mental illness and the abnormalities disturbing daily life and searched for ways of treating them. The rapid development of psychology caught modern man so off guard that he began to attribute additional meanings and implications to this emerging field.

The original task of psychology was to describe reality. Yet, as it developed, psychology began to invade the realm of ethics, and modern man began to confuse the real with the ideal. A new ethic arose that justified human behavior controlled by biological drives. Dr. Akerman, however, sees the "new morality" as simply a continuing rebellion against the hypocrisy and corruption of Christianity.

Dr. Akerman claims that both the Christian and modern outlooks are unhealthy. They both have immoral effects on society and family life. Whereas Christianity unrealistically denies the body, psychology—especially in the more deterministic forms of psychoanalysis and behaviorism—is obsessed with the body and the mechanics of the psyche while neglecting the soul, its Divine image. There must, Dr. Akerman reasons, be a third way out of this dilemma.

## THE THIRD WAY

The third way is the Jewish way. This way unites spirit and body, Heaven and Earth. The healthiness of this system is based on its inclusion of all the components of the human being. Judaism accepts man's drives but refines them. It does not impose ideals of extreme monasticism. The Torah challenges the Jew to elevate and sanctify reality. And just as the results of the non-Jewish outlook are most evident in the deterioration of family life, the fruits of the third way are most pronounced in the flourishing of the Jewish family. The Jewish husband and wife are challenged by *halakhah* (Jewish law) to accept one another as human beings worthy of respect, not just as objects offering sexual satisfaction or a continuation of the "original sin." Impulses and instincts are part of the Jewish family, but instead of drives controlling the marriage partners, the marriage partners try to control their drives.

Modern man reveres the psychoanalyst and the psychiatrist as did primitive man his witch doctor. This veneration should not be total, warns Dr. Akerman. She sees psychology as capable of alleviating only part of human distress. Someone who is satisfied by having his neurosis diagnosed, and at best goes for treatment and tries to deal with his problems, is still not fulfilling all his innate potential. He is treating only the animal part of his soul and not the Divine part.

Nevertheless, in Dr. Akerman's opinion, psychology can play an important and respectable role for Jews today. She explains that during their long exile in Christian countries, European Jews absorbed Christian values without even realizing it. Many Jews internalized the Christian separation of body and soul and adopted the negative Christian stereotype of the Jew as their own self-image.

Many of Dr. Akerman's patients are Holocaust survivors suffering from a low Jewish self-image. But there are also many fortunate Jews who have not directly suffered from the Holocaust or any other persecution but have nevertheless internalized a negative Jewish self-image. Furthermore, Dr. Akerman sees many Jews adopting Gentile attitudes toward Judaism, the most painful example of this being when they adopt anti-Semitic ideas.

In the past, this self-hatred occurred as a result of strong pressure. A Jew like Sigmund Freud did not dare to define his theories in Jewish terms. Although Freud had "Jewish intuition" (in Akerman's words), he formulated his theory of the subconscious by using Greek images and symbols. This way he avoided the massive material on the subject in the Torah, Scripture, and the Talmud. But this way Freud also avoided solving his own identity problem. Compare this to Joseph, who successfully interpreted dreams in a hostile Gentile environment and attributed his success to God (see Genesis 40:8, "Interpretations are God's business," and Genesis 41:16, "It is not by my own power. But God may provide an answer concerning Pharaoh's fortune").

Thus, psychology has a dual role to play for Jews. First, the very science that diverted man's attention from the spiritual ideal to the dust-and-ashes foundation of reality can help the Jew, who is finally physically returning to his land but still hasn't freed himself from the alien values implanted within him during his long exile. A genuine analysis of his soul should lead a Jew to a life of Torah and action. Of course, Dr. Akerman adds, we should not be satisfied with the findings of psychology alone in order to understand the existence and laws of the soul, but we should not refuse to take it into serious consideration either.

Second, psychologists treating assimilated Jews can help them shed anti-Semitic perceptions they have internalized from the general environment. Such Jews are suffering from identity crises because although they are not ready to recognize and appreciate their Jewishness, they are identified by society as Jews. Their Jewish self-hatred is often projected outward against other individuals or groups. This inner conflict splits the personality. Split personalities cannot be cured by illusions. The role of the psychologist here is to help bring such people to their authentic Jewish identity.

## A JEWESS IN GENTILE GARB

Dr. Akerman has much experience in treating such problems. Witness the following case history:

"I received a young female patient who was having problems in all aspects of life. She couldn't find her place in work or in her social life. She was very talented and possessed the qualifications needed for success. When I asked her about her background, I started to understand her problem better. She was born in the Warsaw ghetto in 1942. At the age of six months she was smuggled out of the ghetto and given over to her mother's sister, who was married to a Polish Gentile, living in a village outside Warsaw. She received a Polish name and didn't know she was Jewish until she was sixteen years old. But she didn't know her original name or her parents' name or her birth date. As she was an alert child, she noticed suspicious aspects in the way her aunt and uncle related to her. Her black eyes and hair were a risk for her-self and for the people taking care of her, and she was ordered not to appear in her aunt's courtyard. Questions she asked about the ghetto were never sufficiently answered.

"After the war the girl was taken to a church. There she was taught to pray. Later, she joined the Communist youth movement. There she internalized new identities. Only at age sixteen when she was told of her Jewishness, did she start to understand the meaning of the curses that her uncle gave her aunt when he was drunk and why she had to pretend that she wasn't her aunt's niece.

"After completing high school she went to France, where an uncle lived. But despite her intelligence and talent, she was never happy.

"When she was referred to me, she was already in her thirties. At this age it is difficult to change basic behavioral patterns. It is very hard to question the attitudes upon which the adult personality is built. I tried to help her in several directions. One way in which we worked was the clarification of the past. She searched for people who had known her parents, raking archives in an attempt to produce bits of information about them, about how she was smuggled out of the ghetto. We established contact with survivors of her family in Israel. This was a long process. Every new piece of information, every additional testi-mony demanded immense spiritual effort. Hand in hand with expos-ing the past we examined the present. At this point she decided that she had to leave her Gentile boyfriend. Later she quit her job at the laboratory she ran with him and became independent. After a while, she married a Jew.

"Her problems arose from her internalization of Christian and Communist attitudes toward Jews and thus toward her very self. In addition to all the complications of her life story, this image didn't allow her to live at peace with herself.

"Another course of action was to give her a positive example of Jewish life. Much later the patient told me that the thing which most enabled her to shatter her hostile image of Jewishness was myself—as a proud Jew and mother of a happy family. She identified with me. I was mother, father, grandfather, grandmother, the ghetto, and all the people of Israel for her," Dr. Akerman says, filled with emotion.

"The pictures of the ghetto that she found caused another crisis. It was hard for her to identify with a persecuted, suffering people. When I told her that I was going to participate in a Holocaust survivors' conference in Jerusalem, at first she wanted to go, but in the end she couldn't do it. She feared that a direct confrontation with the Holocaust would take her back to the ghetto (seen in her imagination, in this case, as Israel) and would leave her own children behind in her second past (seen by her as France). She was always trying to escape this dilemma."

Dr. Akerman explains, "The intrusion of a human being into a place he doesn't belong condemns him to failure. He doesn't fulfill himself and he can't successfully realize an identity that isn't his. In the twisted perception of this woman (and not her alone), being Jewish is seen as an inferior status and as lacking an equal right to exist. When she first came to me, she saw a Jew as an intermediary creature between man and the animals. For Jews like her, who survived anti-Semitic persecution, survival itself requires sadistic behavior. People like her see only two possibilities: murderer or victim. If you don't want to be the victim, you must be the murderer. (But not everyone can murder.)

"In my opinion, the Torah suggests a third way, which rescues people from this dilemma. This is the meaning of the vision of Isaiah: 'And the wolf will dwell with the lamb, and the tiger will lie down with the kid, and the calf and the fatling lion cub together, and a little boy will lead them' (Isaiah 11:6). The big will remain big, and the little will remain little, but the strong won't attack the weak. The basic human problem in the case I'm describing is the intrusion of a person into territory not his own. This problem is expressed often on various levels of human relations: between man and wife, parent and child, teacher and student, doctor and patient. Those who suffer or who cause suffering in these relationships don't respect the rights and responsibilities of the other person."

Dr. Akerman continues, "Another crisis for this patient was when she had to change her name to a Jewish one. Her husband's name sounded very Jewish. Taking on this name meant exposing her Jewishness to the world. Her implanted reflex warned that when her Jewishness was known, she would be in danger. Although she had cognitively come to terms with this "threat," it was difficult for her to free herself from her irrational fears. Similarly, she had to struggle with this reflex after she gave birth to her son and faced his circumcision. She had to overcome her fear that cutting her son's foreskin would mark him for danger the rest of his life. Her first annual commemoration of her mother's death was a trying experience, as was the trauma when each of her children reached the age of six months (the age when she was smuggled out of the ghetto). I tried to help her carry on and embrace life, rather than escape and deny her identity."

## MIXED MARRIAGES

Dr. Akerman also describes the identity problems of children of marriages where only one parent is Jewish. She has a more or less standard indicator of their predicament. Such people, she says, tend to lack the strength to get up in the morning. They lack the will to produce the maximum that they are capable of. They don't respect themselves and don't allow others to respect them. They are like singers who try to sing other people's songs in a key not their own. In the end, they don't sing anything. Dr. Akerman tries to direct such people to respect themselves—body and soul—and to accept their past and their heritage. But what exactly is the heritage of the child of a mixed marriage? "If a person has no past, he has no present either," she warns. "And if he has no present, then his future stands in serious doubt."

Mixed marriages are not the direct result of anti-Semitism. But Dr. Akerman sees many similarities between Jews who intermarry and Jews who found themselves forced to pose as Gentiles during the Holocaust. Jews who adopted an entirely materialistic value system in order to survive were also killed by Hitler, says Dr. Akerman. Their death lies in cutting themselves off from the chain of generations.

"When children are ashamed of their parents, their instinct for life is destroyed," says Dr. Akerman. "As parents and educators and psychologists, we are responsible for giving our children a sense of identity and history."

Dr. Akerman places this challenge in a wider perspective. The world expects the Jews to lead the way toward truth, the light of the Torah. But as long as we do not accept ourselves, we cannot serve as an example to anyone. We must weed out the many alien ideas that have overgrown our souls during our long exile. Only after we have realized a healthy and independent identity can we be a model for the world.

From *B'Or Ha'Torah* 7E, 1991. This article first appeared in Hebrew in *B'Or Ha'Torah* 2. It was translated into English by Ilana Coven.

# Notes

## CHAPTER 1

1. A. Gotfryd and R. I. C. Hansell, "The Effect of Observer Bias on Multivariate Analyses of Vegetarian Data," *Oikos* 45 (1985): 223–234; "Prediction of Bird Community Metrics in Urban Woodlots," in *Wildlife 2000: Modeling Habitat Relationships of Terrestrial Vertebrates*, J. Verner, M. L. Morrison, and C. J. Ralph, eds. (Madison, WI: University of Wisconsin Press, 1986), 321–326.

2. I do not intend to ascribe gender to the Divine Being. All terms such as *Hers*, *His*, and *Its* are limiting and therefore ultimately inadequate. These, however, are linguistic limitations, and we can communicate on this topic despite them.

3. P. Davies, "God and the New Physics," *New Scientist* 98 (1983): 872–874.

4. Gotfryd and Hansell, "The Effect of Observer Bias," 223–234.

5. S. K. Langer, *An Introduction to Symbolic Logic* (New York: Dover, 1953).

6. S. A. Kerns, "Who's Playing God?" *Nature* 307, no. 5949 (1984): 312.

7. M. Sachs, "Antiscience Within Science," *The Humanist* 38 (1978): 52–59.

8. M. Gardner, *Fads and Fallacies in the Name of Science* (New York: Dover, 1953); C. N. Klahr, "Science versus Scientism," in *Challenge: Torah Views on Science and Its Problems*, A. Carmell and C. Domb, eds. (Jerusalem: Feldheim, 1978), 288–295; L. Levi, *Torah and Science: Their Interplay in the World Scheme* (Jerusalem: Feldheim, 1983).

9. K. R. Popper, *Conjectures and Refutations* (New York: Basic Books, 1963); *The Logic of Scientific Discovery* (New York: Harper and Row, 1968); G. Schlesinger, *Confirmation and Confirmability* (London: Oxford University Press, 1974).

10. L. von Bertalanffy, *Problems of Life* (New York: Wiley, 1952); E. C. Olson, "Morphology, Paleontology, and Evolution," in *Evolution after Darwin*, vol. 1, S. Tax, ed. (Chicago: University of Chicago Press, 1960), 523–545; M. Goldman, "A Critical Review of Evolution," in *Challenge*, Carmell and Domb, eds., 216–234.

11. L. M. Spetner, "Natural Selection: An Information Transmission Mechanism for Evolution," *Journal of Theoretical Biology* 7 (1964): 412; "Natural Selection versus Gene Uniqueness," *Nature* 226, no. 5249 (1970): 948–949; A. Hasofer, "A Simplified Treatment of Spetner's Natural Selection Model," *Journal of Theoretical Biology* 11 (1966): 338–342; Carmell and Domb, eds. *Challenge*.

12. D. D. Davis, "Comparative Anatomy and the Evolution of Vertebrates," in *Genetics, Paleontology and Evolution*, G. L. Jepson, E. Mayr, and G. G. Simpson, eds. (New York: Atheneum, 1963), 64–89.

13. Klahr, "Science versus Scientism," 288–295.

14. I. W. Knobloch, "Spontaneous Generation," in *Readings in Biological Science*, I. W. Knobloch, ed. (New York: Meredith, 1967), 348.

15. R. J. Blackwell, "Descartes' Laws of Motion," *Isis* 57 (1966): 220–234.

16. G. C. Hatfield, "Force (God) in Descartes' Physics," *Studies in the History and Philosophy of Science* 10 (1979): 121, 126.

17. F. Hoyle and C. Wickramasinghe, *Evolution from Space* (London: Dent, 1981).

18. E. Schrödinger, *What Is Life? and Other Essays* (New York: Doubleday, 1956).

19. A. A. Hirsch, "Scientific Symbols for God," *Chemical Engineering Newsletter* 48 (1970): 10.

20. G. Schlesinger, *Religion and the Scientific Method* (Boston: Reidel, 1977).

21. M. Maimonides, "Book of Knowledge," *Mishneh Torah*, vol 1.

22. W. Blum, "Science for Humanity's Sake," *Science* 64, no. 1647 (1926): 77–80.

23. B. T. Baldwin, "Sigma Xi in Research," *Science* 64, no. 1644 (1926): 6–8.

24. Ibid.

25. P. Morrison, *Powers of Ten* (New York: Scientific American Books, 1982).

26. E. T. Whittaker, *From Euclid to Eddington* (New York: Dover, 1949); D. Bohm, *Causality and Chance in Modern Physics* (Princeton, NJ: Van Nostrand, 1956).

27. E. Wigner, "The Unreasonable Effectiveness of Mathematics in the Natural Sciences," *Communications of Pure and Applied Mathematics* 13, no. 1 (1960): 1-14; W. K. Brooks, *The Foundations of Zoology* (New York: Lemcke, 1967).

28. Hatfield, "Force (God) in Descartes' Physics."

29. Davies, "God and the New Physics," 872-874.

30. G. Holton, "Niels Bohr and the Integrity of Science," *American Scientist* 74 (1986): 237-243.

31. P. Weiss, "The Living System: Determinism Stratified," in *Beyond Reductionism*, A. Koestler and J. R. Smythies, eds. (New York: Macmillan, 1970), 3-55.

32. K. R. Popper and J. R. Eccles, *The Self and Its Brain* (New York: Springer International, 1978).

33. N. L. Rabinovitch, *Probability and Statistical Inference in Ancient and Medieval Jewish Literature* (Toronto: University of Toronto Press, 1973).

34. Spetner, "Natural Selection," 412; "Natural Selection versus Gene Uniqueness," 948-949; Hasofer, "A Simplified Treatment," 338-342; E. H. Simon, "On Gene Creation," in *Challenge*, Carmell and Domb, eds., 208-213.

35. C. S. Elton, *The Ecology of Invasions by Animals and Plants* (London: Methuen, 1958); D. W. Ehrenfeld, *The Arrogance of Humanism* (New York: Oxford University Press, 1978).

36. M. Maimonides, "Book of Judges," *Mishneh Torah*, vol. 13.

## CHAPTER 5

1. E. T. Whittaker, *From Euclid to Eddington* (New York: Dover, 1949).

2. G. Holton, *Introduction to the Concepts and Theories of Physical Science* (Reading, MA: Addison Wesley, 1952).

3. W. von Buddenbrock, *The Senses* (Ann Arbor, MI: University of Michigan Press, 1958).

4. E. Nordenskiöld (Transl. L. B. Eyre). *The History of Biology: A Survey* (New York: Knopf, 1928).

5. E. Schrödinger, *What Is Life? and Other Essays* (New York: Doubleday, 1956).

6. W. Heisenberg, *Physical Principles of Quantum Theory* (Chicago: University of Chicago Press, 1930).

7. Schneur Zalman of Liadi (Transl. N. Mindel). *Likutei Amarim (Tanya)*, vol. 1 (Brooklyn, NY: Kehot, 1969).

8. J. Hadamard, *Lectures on Cauchy's Problem* (New Haven, CT: Yale University Press, 1923; French edition, Paris: Hermann, 1932).

9. S. J. Gould, *Ontogeny and Phylogeny* (Cambridge, MA: Belknap Press of Harvard University, 1977).

10. J. A. Jacobs, *Textbook of Geonomy* (New York: Wiley, 1974).

11. F. Hoyle, "The History of the Earth," *Quarterly Journal of the Royal Astronomical Society* 13 (1972): 41.

12. G. Gamow, *The Birth and Death of the Sun: Stellar Evolution and Subatomic Energy* (New York: Viking, 1940).

13. R. Jastrow, *God and the Astronomers* (New York: Norton, 1978).

14. H. Sperling and M. Simon, transl., *Zohar*, vol. 1 (London: Soncino, 1933), 15a ff, 66 ff.

15. D. D. Davis, "Comparative Anatomy and the Evolution of Vertebrates," in *Genetics, Paleontology and Evolution*, 64–89. G. L. Jepson, E. Mayr, and G. G. Simpson, eds. (New York: Atheneum, 1963).

16. N. Eldredge and J. Cracraft, *Phylogenetic Patterns and the Evolutionary Process* (New York: Columbia University Press, 1980).

17. S. Løvtrup, *Epigenetics: A Treatise on Theoretical Biology* (London: Wiley, 1974).

18. G. W. Burns, *The Science of Genetics*, 5th ed. (New York: Macmillan, 1983).

19. S. Wright, *Evolution and the Genetics of Populations* (Chicago: University of Chicago Press, 1969).

20. Løvtrup, *Epigenetics*.

21. B. Rensch, *Evolution above the Species Level* (New York: Columbia University Press, 1959).

22. M. Maimonides (Transl. M. Friedlander). *Guide for the Perplexed* (New York: Dover, 1956).

23. E. N. da C. Andrade, *An Approach to Modern Physics* (New York: Doubleday, 1956).

24. K. Gödel, "Über format unentscheidbare Säze der Principia Mathematica und verwandber Systeme," I. *Monatshefte für Mathematik und Physik*, vol. 38, 1931, 173–198.

25. S. Toulmin and J. Goodfield, *The Discovery of Time* (New York: Harper and Row, 1965), 18.

26. P. A. M. Dirac, "A New Basis for Cosmology," *Proceedings of the Royal Society*, A 165 (1938): 199.


# CHAPTER 6

## Bibliography

Bohm, D., "Hidden Variables in the Quantum Theory." In *Quantum Theory III: Radiation and High Energy Physics*, D. R. Bates, ed. New York: Academic Press, 1962.

Scott, C., McLaughlin, P., and Chu, F. Y. "The Soliton: A New Concept in Applied Science," in *Procedures of the IEEE*, vol. 61, no. 10 (October 1973): 1443–1483.

Weinberg, S., *Gravitation and Cosmology*. New York: Wiley, 1972.
Zimmerman, C., *Torah and Reason: Insiders and Outsiders of Torah*. Jerusalem: Tvuno, 1979, 108–134.

Negative proof in teleology and many of the arguments of the last section are taken from Ha'Gaon Rabbi Dr. Zimmerman's excellent book.

## CHAPTER 7

1. J. Von Neumann, *Mathematical Foundations of Quantum Mechanics* (Princeton, NJ: Princeton University Press, 1955), esp. notes 207 and 238. Translated from *Mathematische Grundlagen der Quantenmechenic* (Berlin: Springer, 1932, and New York: Dover, 1943).
2. M. Jammer, *The Conceptual Development of Quantum Mechanics* (New York: McGraw-Hill, 1966), esp. 370–373.
3. E. Wigner, "Remarks on the Mind-Body Question," in *Scientist Speculates*, ed. I. J. Good (New York: Basic Books, 1962), 284–302. That an observation performed by a consciousness is also a "measurement" is not disputed. What is noted here is the postulate that *only* a measurement by a consciousness (i.e., an observation) can "collapse" the wave function.
4. J. A. Wheeler, "Genesis and Observership," in *Foundational Problems in the Special Sciences*, R. E. Butts and K. J. Hintikka, eds. (Dordrecht: Reidel, 1977), 3–33.
5. That the wave function is collapsed by measurement is fact, and thus it is part of physics. What causes this collapse is not yet known; various mechanisms are possible. Thus, speculations in this direction are more properly labeled "quantum metaphysics" than quantum physics.
6. All this refers to physical reality from the human perspective rather than the Divine. As Maimonides writes in his *Mishneh Torah* (Moses Hyamson, ed., *The Book of Knowledge* [Jerusalem and New York: Feldheim, 1981], 36a–36b):

All beings, except the Creator, from the highest angelic form to the tiniest insect that is in the interior of the earth, exist by the power of God's essential existence. And as He has self-knowledge, and realizes His greatness, glory and truth, He knows all, and nought is hidden from Him.

The Holy One, blessed be He, realizes His true being, and knows it as it is, not with a knowledge external to Himself, as is our knowledge. For our knowledge and ourselves are separate. But as for the Creator, blessed be He, His knowledge and His life are One, in all aspects, from every point of view, and however we conceive Unity. If the Creator lived as other living creatures live, and His knowledge were external to Himself, there would be a plurality of deities, namely; He himself, His life, and His knowledge. This however is not so. He is One in every aspect, from every angle, and in all ways in which Unity is conceived. Hence the conclusion that God is the One who knows, is known, and is the knowledge (of Himself)—all these being One. This is beyond the power of speech to express, beyond the capacity of the ear to hear, and of the human mind to apprehend clearly. Scripture, accordingly, says "By the life of Pharaoh" and "By the life of thy soul" but not "By the life of the Eternal." The phrase employed is "As God Liveth"; because the Creator and His life are not dual, as is the case with the life of living bodies or of angels. Hence too, God does not apprehend creatures and know them because of them, as we know them, but He knows them because of Himself. Knowing Himself, He knows everything, for everything is attached to Him, in His Being.

7.  J. A. Wheeler, "Beyond the Black Hole," in *Some Strangeness in the Proportion: A Centennial Symposium to Celebrate the Achievements of Albert Einstein*, H. Woolf, ed. (Reading, MA: Addison Wesley, 1980).
8.  A. Rabinowitz, "Free Will," *B'Or Ha'Torah* 6 (1987): 141–158.
9.  *Midrash Bereshit Rabbah* 1:1; *Zohar, Parashat Terumah*, 161b.
10.  Jerusalem Talmud, *Shekalim*, 1: end of *halakhah* 1.
11.  M. C. Luzzatto, *Derekh Ha-Shem*.
12.  *Eruvin* 13b.
13.  Exodus 23:2.
14.  Mishnah, *Rosh Hashanah* 2:9; Talmud, *Rosh Hashanah* 25a.
15.  *Baba Metzia* 59b.
16.  Deuteronomy 30:12.
17.  *Baba Metzia* 86a.
18.  Jerusalem Talmud, *Ketubot* 1:2.
19.  A. Rabinowitz, *And God Said: "Let There Have Been a Big Bang."* In press. This work contains a more comprehensive discussion of the connection between Genesis, evolution theory, the Big Bang, and quantum mechanics.

# CHAPTER 8

1.  The Torah certainly does not present a complete systematic picture of the present structure of the universe nor does it anywhere state that one must believe in some particular geometry of the component parts of the universe (see note 7). What then prompted Christian theologians to construct a so-called biblical cosmology? That they did so is especially surprising since the elements of this cosmology are culled from various brief references scattered throughout the Torah; most of these references are very

marginal to the issue being dealt with and are probably used only as "poetic imagery."

The answer is probably as follows: many pagan customs had been incorporated into Christianity along with the pagans themselves, and so had Greek and other pagan philosophies. Greek logic as formulated by Aristotle reigned in Europe unopposed for many centuries by the time the Church achieved ascendancy there. When the cosmological writings of Aristotle and other ancients were made available to Christian Europe in the twelfth century, they were assumed to be as true as Aristotle's logic and were quite naturally adopted by the Church as well. Just as Scripture was adapted to fit the pagan philosophies regarding theological matters, so too the Greek cosmology was read into the relevant passages. Thus, parallel to the systematic cosmology of the ancient Greeks grew a new "biblical" cosmology.

2. R. Thiel, *And There Was Light* (New York: Mentor, 1962), 123, 125, 145.

3. The attribution of certain contributions to certain of the ancient astronomers is occasionally disputed by various historians. For a critical discussion of the history, see especially Dreyer, e.g., p. 14, note re Anaximander and also re Aristarchus.

4. See reference to Galileo in Hooykaas; see also reference to J. J. Zimmerman in C. Russel, R. Hooykaas, and D. Goodman, eds. *The Conflict Thesis of Cosmology* (Walton Hall, England: Milton Keynes, publisher, 1974).

5. See Rabbi Chaim Zimmerman's *Torah and Reason* (Jerusalem: Tvuno, 1979), 37–38.

6. J. S. Delmedigo, *Elim* (Amsterdam: Menasseh ben Israel, 1629); H. Goodwurm, ed., *Early Acharonim* (New York: ArtScroll, 1989).

7. Historically there were two separate areas of conflict between religion and cosmology: (1) the geometrical structure of the universe as it is at *present*, e.g., geocentrism versus heliocentrism; and (2) the origin and development of the universe from its beginning to its present structure, e.g., Creation versus Big Bang. In this paper we deal exclusively with the first area of conflict.

8. See J. J. Callahan, "The Curvature of Space in a Finite Universe," *Scientific American* (August 1976): 90–100.

9. Of course, in most cases observed by people, moving objects always *do* come to rest unless constantly pushed. However, this is simply due to friction forces between the moving object and the surface it is moving on or the medium it is moving through or due to a collision with another object (which also stops or does not begin to move, due to friction).

10. When one *does* push an object, it resists the motion. This resistance to motion is what is called inertia and it is this resistance which we feel as a "force." The forces which manifest themselves during acceleration (e.g., in a turning car) are a result of inertia and are termed "inertial forces."

11. See note 4.

12. It is an *almost* empty universe because it's empty except for the planet itself.

13. The more mass in the universe, the more inertia. Actually our planet in an otherwise empty universe does have inertia; even a single particle may have inertia in an otherwise empty universe since it can self-interact. However, we do not here consider this very involved, and as yet unsolved, physics dilemma.

14. Motion is undefined because it cannot be observed. It cannot be observed because "motion" of the sole body in the universe causes no observable effects. It causes no effects because the effects of motion are due to inertia, but because the universe is empty there is no inertia. Viewed the other way: empty universe means no inertia means no effects of motion means motion is unobservable, and thus "motion" is a meaningless term in this context.

15. See note 7.

16. See note 8.

17. See K. Gödel, "(An) Example of a New Type of Cosmological Solutions of Einstein's Field Equations of Gravitation," *Reviews of Modern Physics* 21 (1949): 447–450; I. Ozsvath and E. L. Schucking, "The Finite Rotating of the Universe," *Annals of Physics*, vol. 55, no. 1 (London: Academic Press, 1969): 166–204; I. Ozsvath and E. L. Schucking, Letter to the Editor, *Nature* 193 (1962): 1168.

18. See H. Honl and H. Dehnen, "Zur Deutung Einer Von I. Ozsvath und E. Schucking Gefundenen Strengen Losung der Feldgel Eichungen der Gravitation," *Zeitschrift fur Physik* 171 (1963): 178; *Nature* 193 (1963): 362.

19. For a recent symposium discussing the validity of Mach's principles, see *To Fulfill a Vision*.

20. We can describe motion due to gravitational force without using the word *force* by saying that "material particles move along a geodesic of their local space-time, where the local space-time is curved as a result of the matter-energy near and far." (The further away the mass-energy, the weaker its effect.) We don't need the concept of force here. In fact, we need not even say that it is the mass-energy that causes the space-time to curve because, gravitationally speaking, mass-energy can itself be represented as a local concentration of curved space-time. Inventing the name "particle" and so forth to describe this localized curvature is merely convenience.

For example, what we call "the sun" is essentially the sum of its properties as they appear to us. The sum of its gravitational properties can be equivalently termed "a mass" or "a local concentration of space-time curvature."

Therefore, rather than saying, "The Earth and sun attract each other with a mutual gravitational force and orbit a common center of mass," we can say that "the centers of a smaller and a larger space-time curvature exhibit relative motion on geodesics of their local space-time."

21. When the space between the dots doubles, the increase in the distance from any dot to a point two dots away is twice the increase in the distance to a point one dot away. The farther dot moves away twice as fast as the closer dot!

22. Now, of course, when one pictures a balloon or a cake expanding, one thinks of it as taking up more and more of the empty space around it. However, here we wish to talk about the expansion of *space* (space-time) *itself!* Thus, when space-time expands, it does not expand within something, taking up more room; rather it is space itself that expands, and as it expands, it is creating "new space."

23. Only motion in a straight line does *not* involve a change of direction; e. g., motion in a circle involves a constant change of direction.

24. Both speeding up and slowing down are changes in the speed of motion.

25. For example, he gets on a train fifteen minutes early and falls asleep. He wakes up half an hour later while the train is in smooth, uniform motion at eighty miles per hour. However, he believes himself to have slept only five minutes, so that he believes that the train has not yet left the station.

26. Since the train is moving without acceleration, he is neither pressed against his seat, nor pushed from it, nor slipped to the side.

27. And if he sees the stars rotating immediately, then the light from the newly rotating stars has instantaneously reached him.

28. In our case a "local" frame of a star would be a frame of reference affixed to a nearby star.

29. Compare to EPR paradox and Bell's Theorem, etc. See B. d'Espagnat, "Quantum Theory and Reality," *Scientific American* 128 (November 1974): 128–140.

30. The gravitational interaction between the Earth and the sun causes the Earth and the geodesic of space-time around it to have a relative motion. Questions such as "which one really moves?"—the Earth or the geodesic?— are not the concern of physics since only relative motion is meaningful.

As seen from the Earth, the sun will spin on its axis every year, and this can be explained as being due to the fact that the space-time it is centered on is moving in such a way that the relative motion of the Earth and the space-time is a geodesic.

31. From H. Reichenbach, *The Philosophy of Space and Time* (New York: Dover, 1958), 217.

32. Even the contention that the heliocentric description is simpler is not a purely objective description, despite the claim that the heliocentric description would take far fewer bytes (bits of information) than would a description of the geocentric system. As pointed out by Solomonoff, in the words of Robert Wright in the July/August 1987 issue of *The Sciences*:

> The geocentrists, having lost to the heliocentrists by a score of nine billion to nine million bytes, could blame the loss on linguistic discrimination. They could

even invent a computer language that would render the algorithmic description of a heliocentric universe forbiddingly wordy, while permitting quick description of an Earth-centered universe. In defining the language, they would spend millions of bytes describing the idiosyncratic cycles of the planets, then come up with a nice, economical term—say, *fudge*—to refer to that summary. Their algorithmic description of the motion of celestial bodies would consist largely of the statement, "Print fudge."

Of course, this disingenuously designed language would be frowned on by right-thinking scientists everywhere and banished by consensus from serious consideration. But consensus is a subjective thing: there would be no *purely objective*, *purely algorithmic* means of establishing beyond doubt that the geocentrists were playing with a stacked deck. Thus is it always: there is no way to make Ockham's razor fully objective and never will be, according to Solomonoff.

33. From Reichenbach's *The Philosophy of Space and Time*, 219.

34. This was a radical idea, for it was believed until then that Aristotle's proof of the finite spatial extent of the universe was indeed valid. This proof went as follows: If the universe were infinite in extent, at least some of the stars would have to be infinitely distant from the Earth. In a geocentric system, rather than the Earth rotating, all the stars must be considered to complete one revolution every twenty-four hours. In order for the infinitely distant stars to traverse the infinite circumference of the universe in a finite time (twenty-four hours), they must of course travel at an infinite speed. However, since elsewhere in his philosophy Aristotle had "disproved" the possibility of infinite speeds, he had thus "proved" that the universe could not be infinite.

This "disproof" of the infinitude of space was of course based on the "fact" that the stars rotated about the Earth. Thus when Copernicus's theory of the motion of the Earth led to the understanding that there was *no* real rotational motion of the stars, Aristotle's "proof" fell apart. Bruno was the first to have both this insight and the courage to expound it.

35. The universe is known to be vast, possibly infinite. The charge of the insignificance of humanity has been leveled in either case, so the question of the finitude or infinitude of the universe (or that some portions of space-time are beyond our light cone) is irrelevant.

36. From Psalm 8:5–6.

37. Possibly Rambam is referring to people who misunderstand the saying "For me the world was created," in *Sanhedrin* 37a.

38. *Sanhedrin* 37a.

39. Ibid.

40. Rabbi Hisdai Crescas speculated, in the fourteenth century, on the possibility of the existence of many independent universes; and Rabbi Yehudah ben Barzilai discussed the possibility of the existence of intelligent beings on other planets. (See *Sefer Yetzirah* on the question of the "18,000

worlds" mentioned in the Talmud, *Avodah Zarah* 3b. See also N. Rabinovitch's "Conflict or Complement" and N. Lamm's "The Religious Implications of Extraterrestrial Life" in *Challenge*, pp. 44–52 and 354–371.) The question of whether or not there are other intelligent freewilled species in the universe requires a complete volume of its own in order to be adequately discussed. Indeed, a number of scientific books on the subject are in print—but we will merely consider the aspect of it which directly concerns our discussion on the significance of man—and even that only in brief. See I. S. Shklovskii and C. Sagan, *Intelligent Life in the Universe* (New York: Dell, 1966).

41. See Rambam, *Sefer Ha-mada*, chapter 2

42. What the relative significance of humans and alien beings might be in the intellectual and physical sense is impossible to know and really quite irrelevant to the problem at hand.

43. The "area" of a man is about 10 square feet and of a large ranch (10 miles by 10 miles) about 2.5 billion square feet, which is 250 million times larger than the man!

## Bibliography

Armitage, Angus. *The World of Copernicus*. New York: Mentor, 1952.

ben Barzilai, Yehudah. *Commentary on Sefer Yetzirah*. Ed. S. Z. Halberstamm. Berlin: Mekitzei Nirdamim, 1885, 171–173.

Callahan, J. J. "The Curvature of Space in a Finite Universe." *Scientific American* (August 1976): 90–100.

Carmell, Aryeh, and Domb, Cyril, eds. *Challenge: Torah Views on Science and Its Problems*. Jerusalem and New York: Feldheim, 1975.

Crescas, Hisdai. *Or Adonai*. Trans. H. A. Wolfson. Cambridge, MA: Harvard University Press, 1929.

Delmedigo, J. S. *Elim*. Amsterdam: Menasseh ben Israel, 1629.

d'Espagnat, Bernard. "Quantum Theory and Reality." *Scientific American* (November 1979): 128–140. See especially p. 140.

Dreyer, J. L. E. *A History of Astronomy*. New York: Dover, 1953.

Eddington, A. S. *Science and the Unseen World*. New York: Macmillan, 1929, 85–86.

Einstein, Albert. *The Meaning of Relativity*. 3rd ed. Princeton, NJ: Princeton University Press, 1950.

*Encyclopedia of Philosophy*. New York: Macmillan, 1967. S.v. "Copernicus."

Farrington, B. *Greek Science*. Vol. 1. Harmondsworth, England: Penguin, 1944.

Gödel, K. *Reviews of Modern Physics* 21 (1949): 447.

Goldwurm, H., ed. *Early Acharonim*. New York: ArtScroll, 1989.

Grunbaum, Adolph. *Philosophical Problems of Space and Time*. New York: Knopf, 1963.

Honl, H., and Dehnen, H. *Zeitschrift fur Physik* 171 (1963): 178.

Honl, H., and Dehnen, H. Nature 196 (1962): 362.

Hooykaas, R. Religion and the Rise of Modern Science. Grand Rapids, MI: Eerdmans, 1972.

Jammer, M. Concepts of Space. Cambridge, MA: Harvard University Press, 1969.

Lerner, L. S., and Gosselin, E. A. "Giordano Bruno," Scientific American 228 (April 1973) 86–94.

ben Maimon, Moses (Rambam or Maimonides). Guide for the Perplexed. Trans. Shlomo Pines. Chicago: The University of Chicago Press, 1964.

ben Maimon, Moses. Sefer Ha-mada. Vol. 1 of the Mishneh Torah.

Munitz, M. K., ed. Theories of the Universe. New York: Free Press, 1957. See especially the article by J. L. E. Dreyer.

Ne'eman, Yuval, ed. To Fulfill a Vision. Reading, MA: Addison Wesley, 1981. See especially the article by Nathan Rosen and its bibliography.

Ozsvath, I., and Schucking, E. L. "The Finite Rotating Universe." Annals of Physics. New York and London: Academic Press. Vol. 55, no. 1 (October 15, 1969): 55–204.

Ozsvath, I., and Schucking, E. L. Nature 193 (1962): 1168.

Reichenbach, H. The Philosophy of Space and Time. New York: Dover, 1958.

Russell, C., Hooykaas, R., and Goodman, D., eds. The Conflict Thesis of Cosmology. Walton Hall, England: Milton, Keynes Publisher, 1974.

Shklovskii, I. S., and Sagan, C. Intelligent Life in the Universe. New York: Dell, 1966.

Thiel, R. And There Was Light. New York: Mentor, 1962.

Wright, R. Editorial in The Sciences. New York: New York Academy of Sciences. Vol. 27, no. 4 (August 1987): 2–5.

Zimmerman, Chaim. Torah and Reason: Insiders and Outsiders of Torah. Jerusalem: Tvuno, 1979.

For excellent elementary discussion of absolute space
Berry, Michael. Principles of Cosmology and Gravitation. Cambridge, England: Cambridge University Press, 1976.

Davies, P. C. W. Space and Time in the Modern Universe. Cambridge, England: Cambridge University Press, 1977.

For a theory of gravity in conformity with Mach's principle
Sciama, D. W. "Inertia." Scientific American 196 (February 1957): 99–109.

## CHAPTER 9

1. Interestingly, the explosive growth of scientific knowledge in the current era was foreseen in Jewish mystical writings: "And in the sixth century of the sixth millennium the gates of wisdom above and the fountains of wis-

dom below will be opened" (*Zohar* I, 117a [free translation]). "And it appears that this is the reason why now [mid-nineteenth century] there is such an increase in creative developments in secular wisdom, since in the year 5600 [1840 C.E.] the fountains of wisdom were opened" (The Kotzker Rebbe, *Ohel Torah* Section 1).

2. W. James, *The Varieties of Religious Experience* (New York: New American Library, 1958), 71–76.

3. Ibid. 78–79.

4. F. Capra, *The Tao Physics* (Boulder, CO: Shambhala, 1975); J. R. Oppenheimer, *Science and the Common Understanding* (New York: Oxford University Press, 1954), 8–9; W. Heisenberg, *Physics and Philosophy* (New York: Harper & Row, 1958), 202–220; N. Bohr, *Atomic Physics and Human Knowledge* (New York: Wiley, 1958), 20; I. Progoff, *Jung, Synchronicity, and Human Destiny* (New York: Delta, 1971), 67–76.

5. J. O. de LaMettrie, *Man a Machine*, ed. J. Assezata (LaSalle, IL: Open Court Publishing, 1961).

6. D. Bohm. *Quantum Theory* (Englewood Cliffs, NJ: Prentice-Hall, 1951), 26–31; A. Messiah, *Quantum Mechanics*, vol. 1 (New York: Wiley, 1964), 157–159; W. Heisenberg, *Physics and Philosophy*, 44–58; P. A. M. Dirac, *The Principle of Quantum Mechanics*, 4th ed. (London: Oxford University Press 1958), 10–14.

7. E. T. Bell, *Men of Mathematics* (New York: Simon & Schuster, 1965), 181. When Napoleon repeated this to Lagrange, the latter remarked, "Ah, but this is a fine hypothesis. It explains so many things."

8. W. Heitler, *The Quantum Theory of Radiation*, 3rd ed. (London: Oxford University Press, 1954), 16–27; J. Schwinger, ed. *Selected Papers on Quantum Electrodynamics* (New York: Dover, 1958), 209–224; W. E. Lamb and R. C. Retherford, "Fine Structure of the Hydrogen Atom by a Microwave Method," *Physical Review* 72 (1947): 241.

9. Rabbi Menachem Mendel of Lubavitch, *Derekh Mitzvotekha* (New York: Kehot, 1975), 150; Discourses for Shavuot, pp. 139–141.

10. J. Mehra, ed. *The Physicists' Conception of Nature* (Dordrecht, Holland: Reidel, 1965), 244; D. Bohm, & B. Hiley, "On the Intuitive Understanding of Non-locality as Implied by Quantum Theory," *Foundations of Physics* 5 (1957): 96, 102.

11. *Midrash Tanhuma*, "*Pekudei*"; *Avot de R. Natan*, chap. 1; *Zohar* I: 8a, 140a, 250b. II: 20a, 48b, 75b. III: 5b, 117a.

12. Rabbi Schneur Zalman of Liadi, *Torah Or* (New York: Kehot, 1972), 10–11.

## CHAPTER 10

1. Schneur Zalman of Liadi (Transl. N. Mindel). *Likutei Amarin* (Tanya), vol. I, chap. 51 (Brooklyn, NY: Kehot, 1959).

## CHAPTER 11

1. See also A. S. Abraham, *Ha-maayan* 25:25, 1985.
2. *Terumot*, end of chap. 8.
3. *Sefer Meirat Einayim* on *Hoshen Mishpat* 426:2.
4. *Leshonot Ha-Rambam* 1582.
5. *Yehaveh Daat* 3:84.
6. *Pithei Teshuvah* on *Hoshen Mishpat* 426; see also *Nishmat Avraham*, *Yoreh De'ah* 349:2.
7. D. Black, ed., *Renal Diseases* (Oxford: Blackwell Scientific Publications, 1979), 554.
8. F. Vincenti, W. Amend, G. Kaysen, N. Feduska, J. Birnbaum, R. Duca, and O. Salvatierra, "Long Term Renal Function in Kidney Donor," *Transplantation* 36 (1983): 626–629.
9. *Tzitz Eliezer* 10:25(7); *Yehaveh Daat* 3:84; Rabbi S. Z. Auerbach, personal communication.
10. *Minhat Yitzkhak* 6:103.
11. See *Nishmat Avraham*, *Yoreh De'ah* 349:2.
12. *Shevet Me-Yehudah*, p. 314.
13. *Har Tzvi*, *Yoreh De'ah* 277.
14. *Igerot Moshe*, *Yoreh De'ah* 1:229.
15. *Yabia Omer*, *Yoreh De'ah* 3:23.
16. Ibid., 22.
17. *Minhat Yitzkhak* 5:8; *Shevet Ha-levi*. *Yoreh De'ah* 1:211.
18. *Tzitz Eliezer* 14:84
19. Personal communication.
20. *Yoreh De'ah* 339:1.
21. *Yoma* 85a.
22. Ibid., see Rashi.
23. *Hatam Sofer*, *Yoreh De'ah* 338.
24. Maharsham 6:124.
25. *Igerot Moshe*, *Yoreh De'ah* 2:146 and 3:132; *Tzitz Eliezer* 9:46 and 10:25(4). See also *Nishmat Avraham*, *Yoreh De'ah* 339:2.
26. Plum and Posner, *The Diagnosis of Stupor and Coma* (Philadelphia, PA: Davis, 1982); Pallis, "The ABC of Brain Stem Death," *British Medical Journal* 286 (1983): 209.
27. Chief Rabbinate of Israel, "*Hashtalot Lev Be-Yisrael*," *Assia* 11:70 (1987): 70–73.
28. R. Eliashev, personal communication, 1988; R. Vozner, "On Heart Transplantation (Prohibition)," *Assia* 11:70 (1987): 92–94; R. Weiss and R. Waldenberg, *Tzitz Eliezer* 17:66.

## CHAPTER 13

1. Reported in *Nature* 220 (1969): 429.
2. B. Glass, *Science and Ethical Values* (Chapel Hill, NC: University of North Carolina Press, 1965). See also B. Glass, "The Ethical Basis of Science," *Science* 150 (1965): 1254–1261.
3. *Batei Midrashot Vertheimer*, vol. 2, 201, end of *Hashalem*, chap. 5. *Midrash Temurah*.
4. *Evolution: An Analysis from the Jewish Premise* (in manuscript form). The points made in the present article are taken from this book.
5. *Science* 210 (1980): 883–887.

## CHAPTER 14

1. L. M. Spetner, "Information Transmission in Evolution," *IEEE Transactions on Information Theory*, vol. IT-14 (1968): 3–6.
2. H. Bluhme, "Three-Dimensional Crossword Puzzles in Hebrew," *Information and Control* 6 (1963): 306.
3. L. M. Spetner, "Natural Selection versus Gene Uniqueness," *Nature* 226 (1970): 948–949.

## CHAPTER 15

1. A. Carmell and C. Domb, eds. *Challenge: Torah Views on Science and Its Problems* (Jerusalem and New York: Feldheim, 1976), 142–149.
2. M. Ruse, *Darwinism Defended: A Guide to the Evolution Controversies* (Reading, MA: Addison Wesley, 1982).
3. "Evolution" in *Encyclopaedia of Nature and Science* (London: McDonald Co., 1974).
4. F. B. Ford, *Ecological Genetics* (London: Methuen, 1964).
5. L. M. Spetner, "Natural Selection: An Information-Transmission Mechanism for Evolution," *Journal of Theoretical Biology* 7 (1964): 412.
6. A. M. Hasofer, "A Simplified Treatment of Spetner's Natural Selection Model," *Journal of Theoretical Biology* 11 (1966): 338–342.
7. L. M. Spetner, "Information Transmission in Evolution," *IEEE Transactions on Information Theory*, vol. IT-14, no. 1 (January 1968): 1–6.
8. L. M. Spetner, "Natural Selection versus Gene Uniqueness," *Nature* 226 (1970): 948–949.
9. See, for example, the article on "Dating" in the *Encyclopaedia Britannica*, 1974.
10. L. N. Balaam, *Fundamentals of Biometry* (London: Allen and Unwin, 1972), 12.

11. D. V. Lindley, Introduction to Probability and Statistics from a Bayesian Viewpoint, Part I: Probability (Cambridge, UK: Cambridge University Press, 1965), 29.

12. F. B. Salisbury, "Natural Selection and the Complexity of the Gene," Nature 224 (1969): 342–343.

13. See note 3.

14. Ruse, Darwinism Defended.

15. L. M. Spetner, "The Evolutionary Doctrine," B'Or Ha'Torah 2 (1982): 17–25.

16. Carmell and Domb, Challenge, 148.

17. Hasofer, "A Simplified Treatment," 338–342.

18. Salisbury, "Natural Selection," 343, and the references quoted therein.

19. S. P. Champe and S. Benzer, "Reversal of Mutant Phenotypes by 5–Flourouracil: An Approach to Nucleotide Sequences in Messenger RNA," Proceedings of the National Academy of Sciences, Washington, D.C., 48 (1962): 532.

20. Spetner, "Natural Selection," 412.

21. See Salisbury, "Natural Selection," 342–343, and references therein.


CHAPTER 16


1. W. F. Libby, Radiocarbon Dating, 2nd ed. (Chicago: University of Chicago Press, 1955), 33.

2. P. E. Damon, J. C. Lerman, and A. Lang, "Temporal Fluctuation of Atmospheric 14-C: Causal Factors and Implications," Annual Review of Earth and Planetary Sciences 6 (1978): 457–494.

3. I. V. Olsson, ed., "Radiocarbon Variations and Absolute Chronology," in Proceedings of the 12th Nobel Symposium (New York: Wiley, 1970).

4. Damon, Lerman, and Lang, "Temporal Fluctuation," 457–494.

5. M. Rubin and H. E. Suess, "U.S. Geological Survey Radiocarbon Dates II," Science 121 (1955): 481–488.

6. P. P. Tans, A. F. M. de Jong, and W. G. Mook, "Chemical Pretreatment and Radial Flow of $^{14}$C in Tree Rings," Nature 271 (January 1978): 234–235.

7. J. C. Lerman, Les Methodes Quantitatives d'Etude des Variations du Climat au Cours du Pleistocene (Paris: Int., CNRS No. 219), 163–181.

8. Olsson, "Radiocarbon Variations."

9. Damon, Lerman, and Lang, "Temporal Fluctuation," 457–494.

10. Rabbi Yaakov Culi, The Torah Anthology, Yalkut Mei'am Loez, vol. VI (New York: Maznaim, 1979), 145.

11. Rabbi Schneur Zalman of Liadi, Tanya (London: Soncino Press, 1970), Pt. II, chap. 1.

CHAPTER 18

1. A more detailed discussion of the topic may be found in G. N. Schlesinger, *Aspects of Time* (Indianapolis, IN: Hackett, 1980), chapter IV, section 9, "The King, the Nobleman and the Serf."

2. Rabbi Amiel (peace be upon him) is of course not to be held responsible for my free adaptation of his important argument.

CHAPTER 19

1. Y. Breslavi, *Ha-yadata et Ha-aretz*, vol. 6; *"Nof Ha-adam ba-Gallil"* (in Hebrew) (Tel Aviv: Kibbutz Hameuchad, 1964) chap. "Ancient Synagogues in the Gallil," 217–296.

In the Second Temple Period synagogue at Gamla the entrance is located in the southwest side, generally facing Jerusalem, even though the location of its entrance is dictated also by the local topography in which it is built. See S. Gutman, "The Synagogue at Gamla," in *Ancient Synagogues Revealed*, L. I. Levine, ed. (Jerusalem: The Israel Exploration Society, 1981), 30–34. It is structurally more similar to Ein Nashut, a later synagogue built in the Golan. The entrances of the synagogues in Massada and Herodion that were built during the Second Temple Period do not face Jerusalem, but this does not contradict our thesis because their structures were not originally intended to be synagogues.

2. The source for this is in *Tosafot, Berakhot* 6a. In the Talmud (*Berakhot* 6b) there is a discussion about "Whosoever prays at the rear of the synagogue." Rashi explains according to the opinion of the *Tosefta* in *Megillah* cited above that "all synagogue entrances were in the east." The *Tosafot* disagreed, saying *"they* had . . . the custom of praying toward the west . . . but *we* pray to the east."

The Rosh in *Megillah* 3:12 clarified the above statement of the *Tosefta* by explaining, "That was in the days when people prayed toward the west, but we who bow toward the east, toward *Eretz Yisrael*, don't make a doorway in the east."

The Hatam Sofer wrote about this in his *Responsa Orah Hayyim* (responsum 27). He says that this explanation refers to the Talmud Sages in Babylon who prayed toward the west to Jerusalem, whereas the compilers of the *Tosefta* lived in Eretz Yisrael. Their intention was that the synagogue entrance should be in the east wall at any rate with no connection to the direction of praying. The Hatam Sofer says that "serious attention" should be given to the custom of building the synagogue entrance opposite the Holy Ark according to the *Shulhan Arukh*, as opposed to the *Tosefta*. He upholds that the *halakhah* of the *Shulhan Arukh* should be observed and not changed, and he concurs with the Bach that a vestibule should be built in front of a synagogue. The entrance to the vestibule can be on any side, but he prefers the south. My attention was brought to this responsum by Rabbi Menachem Slae of the Responsa Project at Bar Ilan University, and I am grateful to him for it.

3. For further reading, see T. Baras, S. Safrai, Y. Tzafrir, and M. Stern, eds., *Eretz Yisrael from the Destruction of the Second Temple to the Muslim Conquest*, vol. 2 (in Hebrew) (Jerusalem: Yitzchak ben-Tzvi Institute, 1984), 165–189; and Levine (see note 1), articles: N. Avigad, "The 'Galilean' Synagogue and its Predecessors," 42–44; G. Foerster, "Architectural Models of the Greco-Roman Period and the Origin of the 'Galilean' Synagogue," 45–48; E. Netzer, "The Herodian Triclinium—a Prototype for the 'Galilean-Type' Synagogue," 49–51.

4. The Water Drawing Ceremony was such a highlight of the year that the *Mishnah* (*Sukkah* 5:1) remarks that "Anyone who has not seen the Water Drawing Ceremony has not seen joy." It is described also in the Babylonian Talmud, *Sukkah* 51b.

5. See *Mishnah, Yoma* 7:1; *Sotah* 7:7; and in the Babylonian Talmud, *Yoma* 69b, *Sotah* 40b.

6. The structure of the Second Temple Period synagogue at Gamla (see note 1) differs from that described here. It had pillars along all four walls and one central entrance. Its benches flanked the two sides of the entrance, and it didn't have a second floor on top of its pillars. As it was one of a kind, we can't say whether its form was characteristic of the Second Temple Period or whether only after the destruction of the Temple was the structure described here changed in the Gallilean synagogues.

7. See Shmuel Safrai, "Was There a Women's Gallery in the Synagogue of Antiquity?" *Tarbiz* 32:4 (Jerusalem, 1963): 329–338; and also Safrai's book, *Be-shilhei Ha-bayit Ha-sheni Uve-tekufat Ha-Mishnah* (in Hebrew) (Jerusalem: Ministry of Education and Culture, Torah Education Department, 1983), chap. 5, "The Synagogue," 143–163. In Safrai's opinion the gallery was not exclusively for women and there was no *Ezrat Nashim* in ancient synagogues.

On the other hand, see S. D. Goitein, "Women's Galleries in the Synagogues of the Gaonic Period" (eleventh century) in *Tarbiz* 33, 1964, p. 314. Goitein argues that the gallery served as a women's gallery. See also S. Lieberman, *Tosefta Ki-Feshutah* on *Tosefta, Sukkah* (New York: Jewish Theological Seminary of America, 1955), chap. 4, lines 18–19. (The above four references are in Hebrew.)

8. As interpreted by M. Ish-Shalom, in *Beit Talmud*, A.H. Weiss, ed. (Vienna, 1885 and Jerusalem: Carmiel, 1969), 200. See also Lieberman in note 7.

9. See *Yevamot* 105b: "When praying one must put one's eyes down and heart up." Rashi comments: "Eyes down—toward *Eretz Yisrael*." Rabbenu Yonah (*Berakhot*, chap. 5) explains: "One should also think as if one were standing in the Temple which is down." And the *Shulhan Arukh* says (*Hilkhot Tefillah* 95:2) "One should think as if one were standing in the Temple and

in one's heart direct oneself upward to Heaven." And the *Mishnah Brurah* adds that one should feel in one's heart as if one were standing in the Temple in Jerusalem "in the place of the Holy of Holies" (ibid. 94:1:3).

10. See *Baba Kamma* 82a, which attributes to Moses and Ezra the regulation of the number of verses read and the number of readings. On the other hand, the Jerusalem Talmud, *Megillah* 4:1, says that Moses regulated the readings on Sabbaths and holidays, new moon and intermediary festivals, while Ezra regulated the readings on Mondays and Thursdays and Shabbat afternoons. See also *Mekhilta De-Rabbi Ishmael* on *Be-shallah* 1 and tractate *Sofrim* 10, as well as the Jerusalem Talmud, *Megillah* 1:1

11. See S. Safrai, *Be-shilhei Ha-bayit Ha-sheni Uve-tekufat Ha-Mishnah* (see note 7), p. 145. On a Greek inscription from a synagogue dated in the first century B.C.E. found in Jerusalem—the Theodotos inscription—the synagogue is described as having been built "for the sake of reading the Torah and learning the *mitzvot*."

12. The Babylonian Talmud says about the Chamber of Hewn Stone: "It was like a big basilica" *(Yoma* 25a). For a description of the royal portico see Josephus Flavius, *Jewish Antiquities* XV: 411–416. The portico had four rows of columns. Two outside rows were the length of the walls and thus created three spaces. B. Mazar sees this royal portico as no less than the source of the diaspora synagogue because of the synagogue at Sardis. See B. Mazar, "The Royal Portico in the South of the Temple Mount," in *Thirty Years of Archeology in Israel* (Jerusalem: The Israel Exploration Society, 1981), 150, and also *Excavations and Discoveries* (Jerusalem: The Bialik Institute, 1986), 72. However, the synagogue building in Sardis had not originally been constructed to be a synagogue but rather had been part of a pagan gymnasium. The building later served as a civilian basilica until it was given to the Jewish community and transformed into a synagogue. See A. Seager, "The Synagogue of Sardis," in *Qadmoniot* VII: 3–4 (1974): 123–128, and also pp. 178–184 of *Ancient Synagogues Revealed*, edited by Levine (see note 1).

The influence of the royal portico on synagogues in ancient Israel possibly can be attributed to the existence of one on the Temple Mount, such as the Chamber of Hewn Stone. There are those who think that the basilica on the Temple Mount was the court house of the Sanhedrin that sat by the entrance to the Temple Mount (see *Mishnah, Sanhedrin* 11:2), or the place of the Sanhedrin after it was moved out of the Chamber of Hewn Stone. See Babylonian Talmud, *Rosh Hashanah* 31a and *Avodah Zarah* 8b and elsewhere.

## Chronological Order of the Sources and Personalities Quoted in the Text and Notes of Chapter 19

*Mishnah* (plural *mishnayot*): The Oral Law redacted, arranged, and revised by Rabbi Yehudah Ha-nasi at the end of the second century C.E. The word *mishnah* means "learning" and "teaching" and refers to both the six-volume work that Rabbi Yehudah Ha-nasi compiled and the individual articles of law that it contains.

*Tosefta:* Groups of *baraitot* (*mishnayot* that were not included by Rabbi Yehudah Ha-nasi in the *Mishnah*) compiled by *Tanna'im* of the generation before Rabbi Yehudah Ha-nasi, of his same generation and the generation after him. These *baraitot* are a kind of addition and completion of the *Mishnah*. They were edited by the scholars of the generation after Rabbi Yehudah Ha-nasi in the third century C.E. The order of the *Tosefta* is similar to that of the *Mishnah*

*Talmud:* The body of teaching which comprises the commentary and discussions of the *Amora'im* on the *Mishnah* of Rabbi Yehudah Ha-nasi. The study of the *Mishnah* was actively pursued in two centers: *Eretz Yisrael* and Babylon. As a result, two distinct versions of the Talmud emerged: the Jerusalem Talmud (the compilation of which was completed in Tiberias at the beginning of the fifth century C.E.) and the Babylonian Talmud (the compilation of which was completed at the end of the fifth century C.E.).

*Rabbi Yohanan:* An *Eretz Yisraeli Amora* (Talmud Sage) of the second half of the third century C.E.

*Rav Hisda:* A Babylonian *Amora* of the turn of the third century C.E.

*Rashi:* (Rabbi Shlomo Yitzhaki): The greatest Talmud commentator. Lived in France. Died 1105 C.E.

*Tosafot:* Additional comments and interpretations on the Talmud arranged according to the talmudic tractates. The compilers of the *Tosafot* were concentrated mainly in Germany and France in the twelfth and thirteenth centuries. The first compilers of the *Tosafot* were students of Rashi who added notes to his commentaries.

*Rambam* (Rabbi Moshe ben Maimon): Compiled the *Mishneh Torah*, the first comprehensive book of *halakhah* arranged by subject. Died in Egypt in 1205.

*Rabbeinu Yonah* (Rabbi Yonah Gerondi): Wrote commentary on the Talmud and on the halakhic work of the Rif (Rabbi Yitzchak Alfasi of North Africa, who died in Spain in 1103). Died 1264 in Spain.

*Rabbi Meir of Rothenburg:* One of the great sages of Germany. Died in 1293 in prison. (He had been imprisoned by Emperor Rudolf for a huge ransom and instructed the Jewish community not to pay it.)

*Mordekhai* (Rabbi Mordekhai bar Hillel Ha-cohen Ashkenazi): A student of Rabbi Meir of Rothenburg. Wrote a summary of all the *halakhot* from the Talmud up to his times, according to the order of the *halakhot* made by the Rif. Greatly influenced the halakhic rulings after him. Was martyred in 1298 in Nuremburg.

*The Rosh* (Rabbi Asher ben Yehiel): One of the sages of Germany and, later, Spain. Also a disciple of Rabbi Meir of Rothenburg. Wrote a book of *halakhot* according to the order of the Talmud tractates, following the work of the Rif. The rulings of the Rosh together with those of the Rambam and the Rif are the halakhic basis of the *Shulhan Arukh* (see below). Died in Spain in 1327.

*Arba'ah Turim* (or the *Tur*): Book of *halakhah* compiled by Rabbi Ya'akov ben Asher (the son of the Rosh) who arranged all the *halakhot* and customs into four parts (*arba'ah turim*, named after the four rows of precious stones on the breastplate of the High Priest; see Exodus 28:17). Ya'akov ben Asher died in Spain in 1343. The *Tur* was later expounded by two additional works: *Beit Yosef* by Rabbi Yosef Caro, the author of the *Shulhan Arukh*, and the *Bayit Hadash* (or the Bach for short) written by Rabbi Yoel Sirkis (see "the Bach" below).

*Shulhan Arukh*: Compiled by Rabbi Yosef Caro. The *Shulhan Arukh* is the concentration of the halakhic conclusions of the author's *Beit Yosef*, which is based on the *Arba'ah Turim* of Rabbi Ya'akov ben Asher. Caro based his decisions on the Rambam, the Rosh, and the Rif. He died in Tzfat in 1575.

*The Bach*: Acronym of *Bayit Hadash*, the title of the exposition of the *Tur* written by Rabbi Yoel Sirkis, one of the greatest sages of Poland. He died in 1640 in Krakow.

*Magen Avraham*: Commentary on the *Orah Hayyim* part of the *Shulhan Arukh* written by Rabbi Avraham Abely ben Hayyim Ha-levi Gombiner, one of the great sages of Poland. He died there in 1683.

*Hatam Sofer*: Rabbi Moshe Sofer, one of the great rabbis of Hungary in the nineteenth century. Called by the title of one of his books. *Hatam Sofer* (short for *Hiddushei Torat Mosheh Sofer*). Died in Pressburg in 1839.

*Mishnah Brurah*: A clarification of the *halakhot* of the *Orah Hayyim* part of the *Shulhan Arukh* written by Rabbi Yisrael Meir Ha-cohen from Radin, better known as the *Hafetz Hayyim*, after the title of one of his books. He died in 1933 in Lithuania.

# Index

Abortion, 147
Absolute Space, 101–102
  geometric center and, 96–98
  negation of, 99
Abstract logic, *halakhah* and, 230
Acceleration, 109–115
Adam
  and Eve, as an historical
    singularity, 202
  physical reality and, 78
Aesthetic
  as alternative to mundane world,
    264–265
  appeal, scientific theories and,
    208–209
  discreditation of, 262
  need for, 259
Affirmative action, 30
Agassiz, L., on God, 8
Akerman, I., 285–293
  early history, 285–286
  as Jewish role model, 290–291
  on mixed marriages, 292–293
Akiba (Rabbi) on Creationism,
  172–173
Alcalay, R., on *yissurin*, 226
*Almagest*, the, geocentrism and,
  83
Amiel (Rabbi), circular regress and,
  229

Analogous processes, physics and
  biology, 9
Anaximander, geocentrism and, 82
Antiscience, definition of, 5
Archeology
  Alexandrian synagogue, 250
  *Ezrat Yisrael*, 244
  *Ezrat Nashim*, Nicanor Gate and,
    246–247
  facades, synagogues and pagan,
    252–253
  Gallilean synagogues
    *Ezrat Nashim* and, 250
    and Roman basilica, 251–252
  Nicanor Gate, *Ezrat Nashim* and,
    246–247
  synagogue entrances, 242–245,
    253–254
  prayer and
    Babylonian Talmud on,
      238–239
    *halakhah* on, 237
    Jerusalem Talmud on, 239–
      240
    *Midrash Rabbah* on, 240–242
  synagogue structure, *Ezrat
    Nashim* and, 245–253
Aristarchus, heliocentrism and, 82
Aristotelian science, Maimonides
  and, 25

Aristotle
    geocentrism and, 82
    on regularity of nature, 221
Art, mundane world and, 259
Assumptions, scientific, criteria for
    acceptance of, 12
Atheism, Einstein and, 19
Atheists
    explanation of world origin and,
        216
    science and, xiii

Baal teshuvah. See Penitent
Bach, the, on synagogue entrances,
    240, 242
Bartok, B., 281
Beauty
    artistic and scientific, 211
    scientific theory and, 210
    unity and, 211
Behavior, human, scientism and,
    6
Beneficial mutations, natural
    selection and, 6
Big Bang theory
    God and, 213
    Torah and, 42
Biology, analogous processes with
    physics and, 9
Birth and death, concepts of, 170
Blueprint, Torah as
    of Creation, 75
    of the universe, xv
Blum, W., on spirituality, 8
Body, human, control of spirituality
    and, 71
Bohm, D., on the psi functions,
    53
Bohr, N., unity of sciences and,
    9
B'Or Ha' Torah, xvii
Brain
    death, 144
    regulation by the soul, 138

Bribery, belief in evolutionary
    doctrine and, 177–178
Bronowski, J., on religion and
    science, 15
Bruno
    heliocentrism and, 83
    on the infinity of the universe,
        119

Carbon 14 dating. See Radiocarbon
    dating
Cartesian rationalism, 261
Catastrophe theory, 41
Centers, definitions of, 88–94
Chance
    genetic change, probability of,
        179–183
    production of an organized
        system by, 60
Christian
    hypocrisy, identity and, 286
    outlook, modern Humanism and,
        286–287
    split, the modern family and,
        286–287
Christianity, denial of body and,
    287
Church, the
    education and, 17
    political freedom and, 21
    war against, 16
Circular regress, infinite, vicious,
    227–230
Classical physics
    determinism and, 127
    free will and, 127
Communist
    ideology, Ten Commandments
        and, 33
    party, 21
Community health
    definition of, 152–153
    ethical issues in, 145ff.
Comprehension, drive for, 259

Comte's Positivism, 262
Conscious entity, physical reality
    and, 63
Consciousness
    free will and, 69
    measurement and, 67–68
    quantum metaphysics and, 68
Continuous creation, doctrine of, 9
Contrition, heroic quality of, 227
Copernican revolution, the,
    geocentrism and, 81
Copernicus
    heliocentrism and, 83–84
    Jewish writers on, 86
    personal belief in God and, 85
Cosmology
    Big Bang theory, and, 56–57
    historical outline of, 82–83
    Jewish view of, 119–124
    significance of man and, 118–
        124
Court of the Israelites. *See*
    Archeology, *Ezrat Yisrael*
*Creatio ex Nihilo*, Maimonides on, 44
Creation, 34
    continuous
        doctrine of, 9
        renewing of, 70
    historical singularity, as, 201–202
    modern physics and, 57–58
Creationism, Rabbi Akiba's proof
    of, 172–173
Creationists, science curricula and,
    26
Creative inspiration, unity and, 212
Creator, the
    acceptance of, free will and, 58
    acknowledgment of, 3–10
        axioms, assumptions and, 5
        precedents for, 6–7
        theological objections to, 4
    beneficent, 59
    communist ideology and, 33
    ecology and, 10

modern physics and, 4, 58
    purpose of, 61
    reasons for disbelief in, 178
    responsibility to accept, 62
    space–time continuum and, 57–
        58
    teleological argument and, 59–62
    unity of, 12
Cytochrome-C contradiction to
    Darwin's theory, 175

Darwin's theory. *See also* Evolution;
    Evolutionary doctrine; Neo-
    Darwinism
    contradictions of, 175
    cytochrome-C and, 17
    fossil-record gaps and, 175–176
    information transmittal and, 179
    microevolution, macroevolution
        and, 176
    need for quantitative check and,
        174
    qualitative nature of, 174
    random mutation and natural
        selection, 171–172, 187–
        188, 192
    random variation, 173
    refutation of, 183
Dating methods, reliability of, 197,
    200
Davis, D. D., on history of life, 42
Death, talmudic definition of, 143–
    144. *See also* Brain death
Decisions, *halakhah* and, 72–73
Delmedigo, J. S., on Copernicus, 86
Descartes, R.
    acknowledgment of the Creator
        and, 7
    continuous creation and, 9
    on dreaming, 218–219
Determinism
    classical physics and, 127
    Heisenberg's uncertainty
        principle and, 127

Determinism (*continued*)
  human behavior and, 6
  probabilistic, 65
  quantum mechanics and, 127
  universal, xv
Dialectical materialism, xvi
  Torah and, 33
Dirac, P. A. M., on variation of
  natural laws, 44
Divine
  fairness, full pardon and, 225–
    226
  intervention, miracles and, 52
  Mind, divine soul and, 40
  punishment, epidemics and, 28
  revelation, science–religion
    controversy and, 27
  soul, Divine Mind and, 40
  Will
    continuous infusion of, 10
    unity and, 130
Divinity, acknowledgment of,
  precedents for, 6–7
Dreaming, Descartes on, 218–219
Duality, 64–65
  matter-energy, 129
  Divine Will and, 130
  nature of light and, 64
  subject-object, 130–131
  universal, 67

Earth, the
  origin of, Hoyle on, 41
  uniqueness of, 123
Ecological systems, stability of, 10
Ecology
  Creator and, 10
  Seven Laws of Noah and, 10
Eddington, A.
  on heliocentrism *versus*
    geocentrism, 81
  on limitations of knowledge, 37
Education, transfer of from Church
  to state, 17

Einstein, A.
  atheism and, 19
  on a personal God, 60
  on science and religion, 61
  theory of equivalence and, 55
  on unity of social institutions, 15
Emden, J., on Copernicus, 86
Empirical confirmation, rationality
  and, 207, 208
Enlightenment, the, modern
  science and, 261
Environment, outer and inner,
  131–132
Epidemics, Divine punishment, as,
  28
Ether Property, discreditation of,
  97–98
Ethical
  decisions, Jewish, procedure for,
    149
  issues, community health and,
    145ff.
Ethics, Torah and, 149
Eubulides, liar's paradox and, 227–
  228
Euthanasia, 147
Evil Urge, the, 226
Evolution. *See also* Darwin's theory;
    Evolutionary doctrine; Neo-
    Darwinism
  empirical observation and, xiii
  evidence of, 42
  neo-Darwinist school and, 42
  nonscientific nature of, 43
  nontestability of, 43
  organic, information theory and,
    179–183
  prior scientific axiom and, 6
  quantitative analysis and, xiii
  random mutation and natural
    selection, 171–172, 187–
    188, 192
  random variation, 173
  saltationists and, 42

stories of, 42–43
unproven system, as, xiii
Evolutionary doctrine. *See also*
Darwin's theory; Evolution;
Neo-Darwinism
absolute ethics and, 170
accepted phraseology of, 170
belief in, bribery and, 177–178
blind chance and, 169–171
definition of, 170
dominant cultural position of,
169–170
history of, 171
intimidation into acceptance of,
178
Lubavitcher Rebbe on, 185
pre-Darwinian, 171
random mutation, natural
selection and, 171–172,
187–188, 192
random variation, 173
reasons for belief in, 178
role of Creator in, 170
weaknesses in, 169
Evolutionary step, probability of
chance genetic change and,
179–183
Existence of the universe, free will
and, 69
Expanding universe, the
Big Bang theory and, 56–57
closed *versus* unclosed, 56
R(t) and, 56
red shift and, 55
Explanation, science and, 208–
209
External world
independence of, 38–40
proof of, 38
Extrapolation
assumption of uniformity in, 201
backward, 40–41, 44
nature of, 200–201
reliability of, 40

*Ezrat Nashim*
separation of sexes and, 247, 250
synagogue structure and, 245–253
use of during holidays, 249
Water Drawing Ceremony and,
247

Faith
belief in scientific theory and,
43–44
scientific assumptions and, 11
Family life, deterioration of, 287
First Cause, science and, 8
Free will
acceptance of the Creator and, 58
classical physics and, 127
collapse of wave function and,
68–69
consciousness and, 69
decision, definition of, 68–69
existence of the universe and, 69
motivation for the universe, 70
nature and, 71
objective scientific fact and, 71
quantum physics and, 68–69
reality determination and, 78
scientific determinism and, 126–
128
spiritual force and, 127
Torah and, xv
French Revolution, the, 261–262
Freud, S., personal identity
problems of, 288

Galileo
heliocentrism and, 83, 84
personal belief in Bible and, 85
science–religion controversy and,
25
Gamow, G., on Big Bang theory, 42
Gans, D., on Copernicus, 86
General theory of relativity. *See also*
Relativity
cosmology and, 55–56
geocentrism and, 87ff.

General theory of relativity (*cont.*)
    gravitational metric and, 55
    heliocentrism and, 81
    matter-energy duality and, 129
    rationality of, 207
    testing of, 207
    theory of equivalence and, 55
    unity of, 207–208
Genetic change, chance, probability
    of, 179–183
Geocentricity, 94–95, 115–118
Geocentrism, 81, 87ff.
Glass, B., on concepts of good and
    evil, 171
Glucose levels, hormonal regulation
    of, 137–138
God
    alternatives to, 213, 215
    Big Bang theory and, 213
    as explanation of the world,
        213ff.
    fear of, 232
    Master of the infinite universe,
        124
    origin of the world and, 216
    perception of, order of the
        universe and, 58
    rationality and, 207ff.
    search for, 220
    unity with the world and, 222
    universe as an emanation of, 70
Gould, S. J., on catastrophe theory,
    the, 41
Gravitational metric, the, 55
GTR. *See* General theory of
    relativity

Haeckel, E., on evolution, 19
Hajdu, Andre
    on Jewish music, 276–277
    on Jewishness, 279–281
    Judaism and, 272–274
*Halakhah*, 72–79
    abstract logic and, 230

    on bribery, 177–178
    definition of, 74
*Haskalah*, Bible interpretation and,
    xiii
Health
    definition of, 152
    provider, definition of, 152
    services, quality control and,
        159
Heisenberg, W., 9
Heliocentrism, 81–84
Hidden variables, psi functions and,
    53
Hirsch, A. A., acknowledgment of
    the Creator and, 7
Hisda (Rav) on synagogue
    entrances, 237, 238, 239
Homeomorphism, 131
Homeostasis, 135–137
Homosexuality, 147–148
Hormones, homeostasis and, 136
Hoyle, F.
    acknowledgment of the Creator
        and, 7
    on origin of the earth, 41
Human behavior, scientism and, 6
Human mind, mechanistic theories
    of, 127–128
Humanism, modern, Christian
    outlook and, 286–287
Humanities, religion and, 18
Hume, D., on the laws of nature,
    217–218, 221
Huxley, A., on evolution, 19
Hypocrisy, Christian, identity and,
    286
Hypotheses, scientific, falsifiable, 6

Identity, concept and building of,
    286
Industrial Revolution, 261
Inertia, 99–101
Infectious disease, treatment of,
    prayer and, 28

Information theory, organic
evolution and, 179–183
Inspiration, creative, unity and, 212
Intellect, human hunger for, 259

Jackson, Jesse, promise of America
and, 30
Jewish history, science–religion
controversy and, 25ff.
Jewish law
organ transplantation and, 141–
144
reality of the universe and, 70
Jewish medical ethics, 148–150
Jewish mysticism
holistic image of man and, 131
man as internalizer of
environment and, 131
ultimate unity and, 130
Jewish self-hatred, 289
Jewish way, the, 287
Jewish world, the, changes in
twentieth century, 23
Jews
assimilated, 289
internalization of anti-Semitic
perceptions and, 289
science–religion conflict and,
20
Joy and sorrow, simultaneous
experience of, 231–232

Kepler, J.
heliocentric laws of planetary
motion and, 83
on service of God, 15
personal belief in God and, 85
Kerns, S. A., acknowledgment of
the Creator and, 5
Knowledge, limits of, 37–45
Kodaly, Z., 270, 275

Laplace, universal determinism
and, xv

Liar's paradox, 227–228
Libby, W. F., on radiocarbon
dating, 199
Liberal theology, 31–33
Life, history of, Davis on, 42
Light, nature of, duality and, 64
Lockeian environmentalism, 261
Løvtrup, S.
on evolution, testability of, 43
saltationists and, 42
Lubavitcher Rebbe, the, xiv
on evolutionary theory, 185
on neo-Darwinism, popularity of,
193
on physical reality, 63

Mach, E., on inertia, 99, 101
Mahler, G., 276
Maimonides
*Creatio ex Nihilo* and, 44
on direction of prayer, 243
on importance of man, 119
on repentance, three
components of, 234
on science and the Jewish Way,
86–87
on science–religion controversy,
25–26
*Makhshavah*, abstract logic,
*halakhah* and, 230
Man
cosmic significance of, 122
deciding *halakhah* and, 74–75
determination of spiritual reality
and, 74
interpretation of the Torah and,
70–71
as necessary for existence of the
universe, 78
significance of, cosmology and,
118–124
Torah and reality, 70–71
Mankind, reduction of, to animal,
263

Marxism, xvi
Mathematics, unity with physics, 222
Matter, uniformity of, 12
Matter-energy duality, 129
Mayim hayyim (living water),
    analogy of extracellular fluid
    and Torah, 137
Measurement
    physical reality and, 63
    reality and, 66–67
    role of consciousness and, 67–68
Medical ethics
    abortion and, 147
    charitable health organizations
        and, 158
    community consensus and, 146
    consumer and, 150
    euthanasia and, 147
    experiments with human subjects
        and, 162–163
    Gay Caucus and, 147–148
    government as regulator, 160–
        162
    health insurance underwriters
        and, 158
    homosexuality and, 147–148
    individual rights versus
        community responsibility,
        163–164
    interest in, 145–146
    issues, halakhic literature and,
        150–152
    Jewish, 148–150
    National Health Service
        (English) and, 157
    prepaid health plans and, 158
    quality control and, 159
    religious freedom and, 164
    secular world and, 146–148
Medical practice, legal regulation
    of, 147
Medieval period
    mundane struggle and, 261
    spirituality in, 261

Meir (Rabbi) on synagogue
    entrances, 238, 239
Microevolution, macroevolution
    and, 176
Midah ke-neged midah, homeostasis
    and, 136
Milne, E. A., on transitory nature
    of physical theory, 44
Miracles, definition of, 52
Mixed marriages
    children of, poor prognosis for, 292
    deleterious effects of, 292–293
Monotheism, science and, 7–10
Monotheistic tradition, continuous
    creation and, 9
Moses
    fear of God and, 232–233
    science–religion controversy and,
        26
Mundane, world of the
    alternative to, 264–265
    art in, 259
    foci of, 258–260
    moralistic jargon in, 260
    politicization of life in, 260
    relevance in, 259
    World War II generation and,
        262–263
Mutation, beneficial
    improbability of, 191
    Bikini atoll and, 188
    natural selection and, 6
    probability of, 188–193
Mystic revelations, Pascal on, 15
Mysticism–science parallels, 125–126

National Health Service (English),
    ethical issues in, 157
Natural laws
    belief in Torah and, 45
    variation of, 44
Natural selection
    nonscientific nature of theory, 6
    theory of, 171–172

Nature
    free will and, 71
    laws of, 217–218
    materialistic theories of, 19
Nebular Hypothesis, the Torah
        and, 43
Negative Jewish self-image,
        internalization of, 288
Neo-Darwinism. *See also* Darwin's
        theory; Evolution;
        Evolutionary doctrine
    accumulation of small changes
        and, 42
    criteria for scientific explanation
        and, 187–188
    definition of, 186
    popularity of, 193
    random mutation and natural
        selection, 171–172, 187–
        188, 192
    random variation, 173
"New morality," corruption of
        Christianity and, 287
Newton, I., on Creator, 15
Newtonian mechanics, rejection of,
        38

Organ transplantation, 141–144
Origin of the world
    accidental, 223
    atheism and, 216
    God and, 216
Osmosis, homeostasis and, 136–
        137

Pain, 225
Paradox of Einstein, Rosen, and
        Podolsky, 52
Pascal, B., on mystic revelations,
        15
Penitent, profound pain of, 227
Perfection Property, 98–99
Physical reality, 63
    unity of, 129

Physical–spiritual interaction,
        determination of, 71
Physical theory, transitory nature
        of, 44
Physics
    analogous processes with biology
        and, 9
    classical, subject-object relations
        and, 130
    deterministic, xiii
    determinism to indeterminancy,
        51–53
    modern
        acknowledgment of the
            Creator and, 4
        Creation and, 57–58
        subject-object relations and,
            130
    nineteenth century, 51
    unity with mathematics and,
        22
Politicization of life, 260
Popper, K.
    on definition of science, 22
    on scientific truth, xvi
*Posek*, role of, in medical ethics,
        149, 151
Prayer
    infectious disease and, 28
    synagogue entrances and, 237–
        242
Prediction
    backwards, 44
    as primary function of science,
        208–209
Predictive power, scientific theory
        and, 209
Probabilistic determinism, 65
Promise of America, 30
Psychoanalysis, deterministic,
        neglect of soul and, 287
Psychology, 285–293
    role of, for Jews, 288
Ptolemy, geocentrism and, 83

Public health
　characteristics of, 153
　cost of, 157–158
　definition of, 152–153
　expanding role of, 154
　halakhah and, 154
　National Health Service
　　(English), 157
　priorities
　　public policy and, 154–156
　　triage and, 155–156
　scope of, 153
Pythagoras, geocentrism and, 82

Quantum
　electrodynamics, 129
　mechanical tunneling, 54
　mechanics, 127
　metaphysics, 68
　physics
　　halakhah and, 63–79
　　collapse of wave function,
　　　68–69
　　decline of science and, 21
　　epistemological consequences
　　　of, 78–79
　　free will and, 68–69
　　physical reality and, 63
　　probabilistic determinism and,
　　　65
　　two-slit experiment and,
　　　66

Radiocarbon dating
　basic assumptions of, 198–200
　historical singularities, and,
　　202
　methodology of, 198
　sources of error in, 199
　unreliability of, 197
Random mutation and natural
　　selection. See Darwin's theory;
　　Evolution; Evolutionary
　　doctrine; Neo-Darwinism

Randomness
　Divine Will and, 10
　probabilistically determined, 65
Rashi
　explanation of Creation and, 34
　on synagogue entrances, 238
Rationality
　definition of, 212
　empirical confirmation and, 207,
　　208
　general theory of relativity,
　　testing of, and, 207
　God and, 207ff.
　scientism and, 6
　search for God and, 220
　unity and, 208, 212
Ratzo v'shov, definition of, 129–130
Reality
　causation of, 75–77
　determination, free will and, 78
　disclosure of, halakhah and, 73–
　　74
　halakhah, quantum physics and,
　　77–78
　man and Torah and, 70–71
　measurement and, 66–67
　physical, existence of man and, 78
　spiritual realm, the, 69
　Torah and, 69–70
　true, spiritual nature of, 71
Red shift, the, 55
　Big Bang theory and, 57
Redi, F., acknowledgment of the
　　Creator and, 6–7
Relativity. See also General theory
　　of relativity
　geocentricity and, 115–118
　Newtonian mechanics and, 38
Religion
　conflict with
　　science and, 15–23
　　socialism and, 19
　humanities, the, and, 18
　moralistic jargon and, 260

political decline of, 17–18
popular scientific literature and,
18–19
science and, 19, 61
Religious
belief
randomness in nature and, 13
unity of science and, 12
freedom, medical ethics and, 164
Renaissance, rebellion of the
secular against the spiritual,
261
Repentance
contrition and, 227
full pardon, Divine fairness and,
225
governing principle, logical
coherence and, 230
logical intricacies of, 227ff.
love and, 232–234
pain and, 225
paradox of, 231–234
psychological strain of, 227
sin to virtue conversion and, 230
three components of, 234
Romanticism, 262
R(t) definition of, 56
Ruse, M., on natural selection
through random mutation,
187–188, 192
Russell, B., 44
on superiority of science, 20

Saltationists, evolution and, 42
Schneerson, M. M., xiv
Schrödinger, E.
acknowledgment of the Creator, 7
on classical *versus* modern
science, 39
on Divine Mind, the, 40
equations
uncertainty principle and, 52–
53
nonlinear, 53–55

Science
aesthetic appeal and, 12
atheists and, xiii
conflict with religion and, 15–23
decline of, 21
uncertainty principle and, 22
definition of, 5
Popper on, 22
explanation as function of, 208–
209
faith and, 11
falsifiable hypotheses and, 6
First Cause and, 8
foundations of, critique of, 21–22
incomprehensibility of, 21
legitimate religion, as, 19
monotheism and, 7–10
political decline of religion and,
17–18
prediction as function of, 208–
209
primary function of, 208–209
religion and, Einstein on, 61
religious foundations of, 11–13
rise of, 17–18
scientism and, 5–6
self-correction of, 201
spirituality and, 8
superiority of, Russell on, 20
universal validity of, 11
Science–mysticism parallels, 125–
126
Science–religion controversy
contemporary, 26–27
Divine revelation and, 27
immigrant generation, and, 29–
31
Jewish history and, 25ff.
modern religious establishment
and, 31–33
Moses and, 26
motivations for, 34–35
self-preservation and, 29
Soviet Russia and, 33–34

Sciences, unity of, 9
Scientific
    assumptions, criteria for
        acceptance of, 12
    determinism, free will and, 126–
        128
    explanation, criteria for, 186–
        187
    instruments, use of, 38–39
    laws, uniformity of, 12
    literature, effect on religion, 18–
        19
    method, the, definition of, 38
    observation, use of instruments
        in, 38–39
    theories, contradictory, selection
        of, 209
    theory
        aesthetic appeal of, 209
        beauty and, 210, 211
        criteria for acceptance of,
            208–209
        disproof of, 5–6
        faith and, 43–44
        proof of, 38–39
        simplicity and, 210, 211, 212
        sufficient conditions for
            acceptance of, 209–210
        unity of, 209
        working hypothesis and, 43
    truth, xvi
Scientism, 5–6
Scientists, religious, role of, 23
Seven Laws of Noah, the, ecology
    and, 10
Shimon (Rabbi) on *yissurin*, 226
*Shulhan Arukh* on synagogue
    entrances, position of, 238–
    239, 243
Simple, definition of, 213
Simplicity
    definition of, 210–211
    scientific theory and, 210
Sinai, historical singularity, as, 202

Singularity, historical
    Adam and Eve, as, 202
    Creation, as, 201–202
    radiocarbon dating and, 202
    Sinai, as, 202
    Torah on, 201–203
Social institutions, unity of, 15
Socialism, conflict with religion
    and, 19
Solitons, 54
Soul, the, as regulator of the brain,
    138
Space-time
    continuum
        Creator and, 57–58
        expanding universe and, 56
    expansion, 104–106
Spiritual
    control of, human body and, 71
    force, free will and, 127
    fusion with physical, Torah and,
        71
    physical interaction
        determination of, 71
    reality, determination by man of,
        74, 77
Spirituality, science and, 8
Subject-object relationships,
    experimental science and,
    130–131
Sufficient reason, principle of
    applications of, 214–215
    origin of the world and, 213
Synagogue, function of, 251

Technology, negative effects of, 22
Ten Commandments, communist
    ideology and, 33
*Teshuvah. See* Repentance
*Teshuvah me'ahavah. See*
    Repentance, love and
Theology, liberal, 31–33
Theory of equivalence, general
    theory of relativity and, 55

Torah, the
  Big Bang theory and, 42
  conflict with science, xiv
  Creation , blueprint for, 75
  dialectical materialism and, 33
  ethics and, 149
  extracellular fluid analogy, 137
  free will and, xv
  fusion of spiritual and physical
    and, 71
  heliocentrism and, 86
  historical singularities and, 201–
    203
  interpretation of, 70–71
  man and reality, 70–71
  meaning and reality and, 69–70
  mundane world alternative, 264
  nature of existence and, 126
  Nebular Hypothesis, the, and, 43
  public reading of, 251
  rewriting of, 31–33
  science and, xiv, 32
  secular Western worlds and,
    257–258
  synthesis with science, 32
  universe, blueprint for, xv, 70
Tree-ring dating, reliability of, 197,
  200
Triage, 155

Uncertainty principle, 9, 39–40
  decline of science and, 22
  deterministic physics and, 127
  hidden variables and, 53
  philosophical extension of, 54–
    55
  quantum mechanical tunneling
    and, 54
  Schrödinger equation and, 52
Uniform motion
  *versus* accelerated motion, 106–
    107
  relativity of, 107–109
Uniformitarian principle, the, 41

Uniformity of matter, 12
Unity
  argument from, 217ff.
  beauty and, 211
  complexity and, 211–212
  Creation and, 12
  creative inspiration and, 212
  general theory of relativity and,
    207–208
  God and, 222
  mathematics and physics and,
    222
  natural law and, 221
  nature of reason and, 221
  rationality and, 208
  scientific theories and, 208–209
  search for God and, 220
  ultimate, Divine Will and, 130
  understanding and, 211–212
Universal determinism. *See*
    Determinism, universal
Universe, the
  God, perception of and, 58
  motivation for, 70
  order of, 58
  consciousness and, 68
  emanation of God, as an, 70
  expanding, 102–104
  geometric center of, 95–96
    Earth and, 102
  Jewish
    law and, 70
    view of, 119–124
  man as necessary for the
    existence of, 78
  origin of, theories of, 41–42
*Uvkein*, Jewish mysticism and, 131–
  132

Valued acquisitions, payment for,
  226
Von Neumann, J., on consciousness
  and collapse of the wave
  function, 67

Wasserman, E., on bribery, 177–
178
Water, beneficent Creator and, 59
Water Drawing Ceremony, Ezrat
Nashim and, 247, 250
Wave function, collapse of, 66, 68–
69
Wheeler, J., on consciousness and
the universe, 68
Wigner, E., on quantum
description, consciousness
and, 68
Women's Court. See Ezrat Nashim
Working hypothesis, scientific
theory and, 43
World, the
age of, Torah versus scientific,
197

Health Organization, 152
origin of
God and, 216
persistence of question,
216
principle of sufficient reason
and, 213
unity of, 217
with God, 222
Western, secular, degeneration
of, 257–258
World Health Organization,
152
Wright, S., on evolution, 43

Yissurin, 226
Yona (Rabbeinu) on synagogue
entrances, 238

## About the Editors

A well-known educator in the fields of the philosophy of science and Jewish religion and culture, Herman Branover is professor and head of the Center for Magnetohydrodynamic Studies at Ben Gurion University of the Negev. A former Soviet refusenik, Professor Branover is dedicated to bringing the teachings of Judaism to Jews in Russia as well as the rest of the world. While still in Russia, despite being frequently arrested, interrogated, and harassed, he personally taught Jewish ethics to many individuals and groups. He also began to translate Jewish texts into Russian, an endeavor he continues in Israel under the auspices of SHAMIR, an association of religious academic immigrants from Russia in Israel, of which he is the chairman. SHAMIR has published more than 300 titles in Israel and Russia. The holder of a Ph.D. and a Doctor of Science degree, Professor Branover has published many books on both Judaism and science. In collaboration with the Russian Academy of Science, he is currently at work on a five-volume encyclopedia, *Jews in Russia*, covering the 500-year history of Jews in Russia and their contributions to the country, particularly in the sciences. He is the founder, publisher, and editor-in-chief of the Hebrew-English magazine *B'Or Ha'Torah*.

Ilana Attia is the managing and literary editor of *B'Or Ha'Torah*. She holds a bachelor's degree in English literature from the University of Massachusetts and was the translator of Herman Branover's work *Return*. She lives with her husband in Jerusalem.